CÔTE D'OR

THE CLASSIC WINE LIBRARY

Editorial board: Sarah Jane Evans MW,
Richard Mayson and James Tidwell MS

There is something uniquely satisfying about a good wine book, preferably read with a glass of the said wine in hand. The Classic Wine Library is a series of wine books written by authors who are both knowledgeable and passionate about their subject. Each title in The Classic Wine Library covers a wine region, country or type and together the books are designed to form a comprehensive guide to the world of wine as well as an enjoyable read, appealing to wine professionals, wine lovers, tourists, armchair travellers and wine trade students alike.

Other books in the series include:
Port and the Douro, Richard Mayson
Sherry, Julian Jeffs
The story of champagne, Nicholas Faith
The wines of Australia, Mark Davidson
The wines of California, Elaine Chukan Brown
The wines of central and southern Spain, Sarah Jane Evans MW
The wines of Chablis and the Grand Auxerrois, Rosemary George MW
The wines of Germany, Anne Krebiehl MW
The wines of Greece, Konstantinos Lazarakis MW
The wines of New Zealand, Rebecca Gibb MW
The wines of northern Spain, Sarah Jane Evans MW
The wines of Piemonte, David Way
The wines of Portugal, Richard Mayson
The wines of Roussillon, Rosemary George MW
The wines of South Africa, Jim Clarke
Wines of the Languedoc, Rosemary George MW
Wines of the Loire Valley, Beverley Blanning MW
Wines of the Rhône, Matt Walls

CÔTE D'OR

The wines and winemakers of the
heart of Burgundy

Second edition

RAYMOND BLAKE

ACADEMIE DU VIN LIBRARY

Raymond Blake is one of Ireland's leading wine writers. His enthusiasm for wine is boundless and as an independent voice his judgement and opinions are widely respected. He writes the Burgundy entries for the annual *Hugh Johnson's Pocket Wine Book* and is a regular contributor to numerous magazines including *The World of Fine Wine* and *Sommelier India*. He also writes for wine-searcher.com. His travels have taken him to most of the far-flung corners of the wine world, though his spiritual home is Burgundy, where he regularly leads bespoke tours for wine lovers from across the globe. He is also much in demand as an organizer and speaker at themed wine dinners. He is a member of the Circle of Wine Writers and in 2006 was inducted as a Chevalier du Tastevin in Burgundy – at the same time as he and his wife bought a house in Côte d'Or. His other wine books include *Breakfast in Burgundy* and *Wine Talk* (see bibliography).

For Fionnuala

Copyright © Raymond Blake, 2017, 2025

The right of Raymond Blake to be identified as the author of this book has been asserted in accordance with the Copyright, Designs and Patents Act 1988.

First published in 2017 by Infinite Ideas Limited

This edition published 2025 by Académie du Vin Library Ltd
academieduvinlibrary.com

All rights reserved. No part of this publication may be reproduced, stored in a retrieval system, distributed or transmitted in any form or by any means, including photocopying, recording or other electronic or mechanical methods, and including training for generative artificial intelligence (AI), without the prior written permission of the publisher. Translation into any other language, and subsequent distribution as above, is also expressly forbidden, without prior written permission of the author and publisher.

A CIP catalogue record for this book is available from the British Library
ISBN 978–1–913141–94–3

Brand and product names are trademarks or registered trademarks of their respective owners.

Front cover: © Mick Rock/Cephas Picture Library.
Back cover: tichr/shutterstock.com

Photos: page 3 courtesy of F. Hunt, page 280 courtesy of Domaine Chartron, page 166 courtesy of Michel Joly. All other photos courtesy of the author.

Maps, pages 65, 74, 92, 113, 122, 143, 158, 181, 184, 197, 217, 226, redrawn by Darren Lingard courtesy of BIVB (www.vins-bourgogne.fr) from bourgogne-maps.fr

Publisher: Hermione Ireland
Editor: Rebecca Clare
Indexer: Catherine Hall

Typeset by Suntec, India
Printed in Great Britain

CONTENTS

Acknowledgements		ix
Introduction		1
1	A brief history to 1985	5
2	Vineyards and viticulture	18
3	Producing burgundy	47
4	The villages and producers of the Côte de Nuits	64
5	The villages and producers of the Côte de Beaune	157
6	Recent vintages – enjoying burgundy	256
7	The Côte d'Or today	275
8	Visiting the Côte d'Or	289
Glossary		298
Bibliography		301
Index		305

MAPS

Map 1: The Côte de Nuits	65
Map 2: Gevrey-Chambertin	74
Map 3: Morey-Saint-Denis	92
Map 4: Chambolle-Musigny and Vougeot	113
Map 5: Vosnée-Romanée and Flagey-Echézeaux	122
Map 6: Nuits-Saint-Georges	143
Map 7: The Côte de Beaune	158
Map 8: Pommard	181
Map 9: Volnay	184
Map 10: Meursault	197
Map 11: Puligny-Montrachet	217
Map 12: Chassagne-Montrachet	226

ACKNOWLEDGEMENTS

As with all my wine books not a single word would have made it onto the pages that follow without the unfailing, patient and endlessly encouraging support of my wife, Fionnuala. She was with me every step of the way and no words are adequate to describe the background role she played to see this work to fruition. Though I did the legwork and the writing, in many respects this book is a joint effort. Fionnuala was an invaluable collaborator whose calm counsel I could not have done without. Her belief kept the wheels on when the task seemed insurmountable.

Many other people helped in numerous other ways. I am greatly indebted to all the vignerons of the Côte d'Or who opened their doors to me and gave of their time, not just to taste wine but to talk Burgundy past, present and future. The knowledge and understanding thus gained were invaluable. Their names are too numerous to list here but a producer profile of each will be found in the pages that follow. Their dedication to the task of producing the best wine they can in challenging and varying circumstances is humbling and inspirational. For facilitating many dozens of visits to these vignerons I owe a special debt of gratitude to Cécile Mathiaud and Emilie Murat of the Bureau Interprofessionnel des Vins de Bourgogne (BIVB) who gave me unstinting assistance at what can only be described as *grand cru* level.

A particular word of thanks is due to Jean-Marc Vincent and his wife Anne-Marie, with whom some fine bottles have been shared and insights gained. Jean-Marc is one of the most thoughtful and articulate of Côte d'Or winemakers, who always has something interesting to say on any aspect of winemaking or viticulture. Aubert de Villaine and his wife Pamela are similarly gifted and regularly gave of their time to answer my innumerable questions. Walking the vineyards with Aubert was a particularly enlightening exercise. Many exhausting but stimulating days were spent traversing the length and breadth of the Côte d'Or researching this book and I am especially grateful to my neighbours Joël Provence and Patricia Bouchey who, knowing my schedule, frequently invited me for dinner and excellent wine at the day's end. It was a gesture for which no words of gratitude are adequate.

Many others, family, friends and colleagues, offered assistance as well as words of encouragement and, in many cases, shared precious old bottles: Jean-Claude Bernard, Pat Blake, the Boisset family, David Browne, the late Tomás Clancy, Brian Doran, Prof. John Gaskin, John Glen, Juris Grinbergs, Michael and Kate Hayes, Ivan Healy, Dougie Heather, Andrew Keaveney, Bill Kelly, Noel Kierans, Des Lamph, Aiden Maguire, Andrew and Sue Mead, Donal Morris, Monica Murphy, Micheál O'Connell, James O'Connor, Patrick O'Connor, Mark O'Mahony, the late Jim Pilkington, Lochlann and Brenda Quinn, Margaret Rand, Jo Rodin, the Seysses family, Dómhnal Slattery, Simon Spence, Steven Spurrier, Dr Mahendra Varma, Nigel Werner, David Whelehan, the late Sir Brian Williamson and Michael Yon; as well as my parents Gay and Frank, and my sisters Barbara and Margaret. And special thanks to *les aviateurs*, Jim Tunney and Brian Spelman, who flew me up and down the côte to gain a bird's eye view of the famed vineyards and villages. Thanks also to the team at Academie du Vin Library, Rebecca Clare and Hermione Ireland, as well as the editorial board of Richard Mayson, Sarah Jane Evans MW and James Tidwell MS who lent support and advice at critical junctures as the manuscript took shape. To any whom I have inadvertently overlooked I offer my sincere apologies and the promise of a good glass of burgundy when we next meet.

Finally, a special and sad word of gratitude to my late wine-loving friend Eddie O'Connor. 'Eduardo' was an inspirational force to all who knew him and in my case his energy, enthusiasm and support drove the project on in the early days, when it might otherwise have stalled. His support, encouragement and advice over many years proved valuable in more ways than these few words can ever express. I remain forever in his debt. Rest in Peace, Eddie.

INTRODUCTION

The Côte d'Or, the Golden Slope, enjoys a reputation and exerts an influence in the wine world out of all proportion to its size. It possesses none of the grandeur of the Douro Valley, for instance, nor the picture-postcard beauty of South Africa's winelands. It is not majestic; its beauty is serene, and what strikes the observer time and again is the tiny scale. From north to south it is about 50 kilometres long, is sometimes less than a kilometre wide, and can be driven in little more than an hour. At a push it could be walked in a day. Yet for a thousand or more years this favoured slope has yielded wines that have entranced and delighted wine lovers, with a fair measure of frustration and disappointment thrown in too.

Avoiding hyperbole when writing about the Côte d'Or is a problem, for it has held people in thrall for centuries, frequently prompting poetic flights of prose: 'The Pathetic Fallacy resounds in all our praise of wine ... This Burgundy seemed to me, then, serene and triumphant ... it whispered faintly, but in the same lapidary phrase, the same words of hope.' Thus wrote Evelyn Waugh in *Brideshead Revisited* eighty years ago. Literary critics might cavil at such prolixity, and Waugh himself tamed it severely in later editions, but it was enough to intrigue me decades ago and engender a love for the region and its wines that has never waned.

Greater Burgundy is a bigger, more geographically diverse, region than might first be supposed, stretching from Chablis, south-east of Paris, to Beaujolais, north of Lyon. The subject matter of this book, however, is Burgundy's heart, the Côte d'Or. There is no more celebrated stretch of agricultural land on earth. It has been pored over and analysed, feted and cosseted, obsessed about and sought after for centuries and today, if anything, it exerts a greater pull on wine lovers than ever before. 'Astronomical' does not begin to describe the prices now being paid for prized patches of *grand* and *premier cru* vineyard, and for the wines produced from them by the top domaines. Such prices hog the headlines and paint a dazzling, though severely one-dimensional, picture of the Côte d'Or today. In the early years of the twenty-first century they are part of the story but they are not the only part.

This book aims to get behind the dazzle and flesh out the story, to add some nuance and extra dimension. The Côte d'Or cannot be captured in a sound bite nor, it must be admitted, in a book of this scope – but my hope is that it may add another chapter to the ever-unfolding tale, setting it in the early years of the twenty-first century, a period that may come to be considered by future historians as a golden age for Burgundy but which has brought its own challenges in the shape of those ludicrous prices, the scandal of premature oxidation in the white wines, and the increasing challenge of dealing with extreme weather events such as hail and spring frost.

The Côte d'Or has been the subject of forensic scrutiny for centuries, generating a library of books, so why another one now? For the simple reason that it is ever changing. Every year sees new names added to the producers' roster and old ones slipping away, and thanks to this ongoing evolution the infant domaine of today can be the superstar of tomorrow. The core of the book comprises about a hundred producer profiles. Many fine domaines and négociants whose wines I am happy to purchase and drink have not found a place here. It is important to stress that they have not been included, rather than excluded, simply for reasons of space. The aim was to feature a representative collection of producers, not a top-down selection of the most celebrated names in descending order of renown. As a consequence this book is not, nor was it ever intended to be, a comprehensive A to Z of producers.

I write as an *amateur du vin*, an enthusiast whose admiration for the wines of Burgundy stretches back over four decades, and who has been visiting the region and writing about the wines for some thirty years. I do not trade in wine. I try to see beyond the wine to get something of the backstory, the story of people and place that makes the Côte d'Or so fascinating. To examine the wines in isolation is to dislocate them from that, and not knowing something of the backstory precludes a full understanding of them. A broad lens must be brought to bear on the

The author collecting wine by wheelbarrow

côte. Too narrow, and the wood will never be seen for the trees. I have learned much by coming across things serendipitously rather than dashing hither and thither, seeing a lot but noticing little. Without time for assimilation and reflection, subtlety and shade are missed.

Finally, no gustatory experience can match the thrill of a great Côte d'Or red drunk at its peak. The colour, crimsoned by age; the heavenly scent, perfumed with notes of sweet decay; a *sauvage* edge, the palate lively and tingling, managing to be so many things at once, oscillating between fruit and spice and meat and game, a merry-go-round of flavour, spiralling on the palate, refusing to be pinned down by anything so prosaic as a tasting note. All the primary components melded by age and yielding up new ones, unsignalled when the wine was young. Everything cohesive and in harmony, like a great orchestra playing at its best. Above all, vital and living, endlessly enchanting and intriguing, engaging the palate and the spirit like no other wine.

SOME NOTES

The terms 'village' and 'commune' are sometimes used interchangeably. In general I use 'commune' to indicate the vineyard area surrounding a village, and 'village' for the urban heart of the commune, but they tend

4 CÔTE D'OR

to overlap and there isn't a rigid distinction between them. In common parlance, village is more widely heard than commune. When written in italics, *village* is used to indicate the rank of a vineyard and its wine, so that in the hierarchy of vineyard classification *village* comes below *premier cru* and *grand cru*. A wine labelled simply Gevrey-Chambertin or Chambolle-Musigny is a *village* wine.

The Côte d'Or runs in a south-south-west direction from Dijon but for simplicity's sake when, for example, describing the relative positions of different villages to one another I use the cardinal compass points. Thus Pommard is 'north' of Volnay and Vosne-Romanée is 'south' of Vougeot. The same applies to east and west. Greater accuracy is employed when mentioning the orientation of a specific vineyard or slope.

Each producer profile includes a 'try this' note about one of their wines. It could be their greatest wine, their simplest one, or something in between; the criteria for selection were loose and purely personal. Each stands as an individual, and should be seen as such: the wines do not form a homogeneous group, nor is it a parade of flagship wines. The wines represent the house style and ethos of each producer and in many cases they punch above their weight in terms of price or appellation. 'Try this' is not a formal tasting note, it is meant to highlight distinctive and characterful wines that I believe are worth seeking out.

Note to the second edition

Almost all the material in the producer profiles in Chapters 4 and 5 is new, while other sections of this second edition have been updated and modified where necessary. Thus the profiles form the core of the book, in each case giving a snapshot of a domaine's or a négociant's philosophy and practices. In describing how scores of winemakers go about their work the risk is that the profiles can read repetitively so, to combat this, I have sought to highlight distinctive features that mark out one winemaker from another, what sets each apart from the other. Because so much is new in Chapters 4 and 5 this second edition can be used as a companion volume to the first.

1

A BRIEF HISTORY TO 1985

Beneath the streets of Beaune, in the cellars of Joseph Drouhin, the twelfth-century section dubbed the 'Cellar of the Kings of France' is built on the foundations of a Roman fort. Nearby is *La Collégiale*, a cellar that dates from the thirteenth century and which is classified as a historical monument. It is built above the source of a stream named Belena, from which the name Beaune is derived. More importantly for the citizens, it was the original source of their drinking water. History runs deep in Burgundy's Côte d'Or.

The vine has been planted in the Burgundy region, if not specifically on the narrow hillside strip that constitutes the Côte d'Or, for about two millennia. Vine cultivation in Roman times was widespread but probably scattered. It shared the land with other forms of agriculture and the blanket monoculture of today, with the vines intensively tended and trained, had yet to take hold. What is worth noting is that the vineyards were planted on flat ground that was not well drained. Ease of tending probably prompted this, with quantity rather than quality as the ultimate goal.

A much clearer picture of Roman viticulture emerged in 2008 when an ancient vineyard dating to the first century AD was discovered near Gevrey-Chambertin. It covered an area of about 3 hectares with rows arranged in a regular, carefully measured pattern. Hollows were dug in each row, with a pair of vines planted in each, separated by stones so that their roots did not become entwined. It appears the vines were propagated by *provignage* or layering, whereby a shoot of the vine was bent and buried in the soil, there to take root, a method practised up until the time of phylloxera in the late nineteenth century. Later, in

AD 312, written evidence of vine cultivation is found in a submission to the emperor Constantine pleading for fiscal leniency by way of reduced taxes. To back up the plea, the submission detailed a baleful litany of decline and adversity, outlining the problems faced by the wine growers. Drainage channels were blocked through neglect, rendering the good land swampy, the vines were untended and the vineyards chaotic. The region of Arebrignus, today's côte, was in a sorry state, abandoned in parts where it was populated by wild animals. Even allowing for a certain gilding of the lily to soften the emperor's heart, these travails help to put today's problems of frost, hail and the like into a more tolerable perspective.

It is not clear exactly when vines began to be planted on the hillsides of the Côte d'Or but the regimented symmetry of today's *vignobles* with their ordered ranks of arrow-straight rows was still far in the future. Certainly, by the sixth century the vineyards had begun their creep up and away from the flat lands of the plain into less fertile areas. It was a logical move to plant vines where they could thrive and where cultivation of other crops would meet with poor results; land more suited to them was also freed up by this process.

The name 'Burgundy' comes from the Burgondes, a people who moved westwards into the region from Germany in the fifth century. Apart from giving their name to the region, blame might also be laid at the Burgondes' door for starting the process of regulation that has developed into the labyrinthine, bureaucrat's dream of today. To be fair, they were only codifying the law, stipulating what was and was not permitted in the vineyards, and not concocting a Byzantine nomenclature. As the Romans abandoned their lands due to a shortage of labour to work them, the Burgondes moved in to plant them with vines. The law in this respect was clear: if the legal owner did not immediately object then the newcomer only had to compensate him by way of gifting him another piece of land equal in area to the newly planted vineyard. If, however, the land was planted against the owner's wishes then he could claim the new vineyard as his.

In 630 the abbey of Nôtre Dame de Bèze was founded by Duke Amalgaire who granted the monks a sizeable area of vines in Gevrey and elsewhere. Their memory is preserved in the Clos de Bèze vineyard name, which, along with many others, acts as a historical marker in the story of the Côte d'Or. Indeed, a study of vineyard and place names, tracing their origins back over the centuries, makes for a revealing if challenging investigation of the côte's history. Corton-Charlemagne is the most resonant of all, recalling that the Holy Roman Emperor owned vines on the hill of Corton, reputedly having ordered they be planted there when he noticed winter snow melting earlier on the hill than elsewhere.

Moving towards the end of the first millennium, the foundation of the Benedictine abbey at Cluny in 910 could justifiably be regarded as the most significant date in Burgundy's history. In time the abbey grew to be the largest Christian building in the world and despite the depredations it suffered after the French Revolution, when it was used as a handy source of stone for building, it is still worth visiting to gain an appreciation of its scale. Thanks to grants and donations of prime land from local lords, noblemen and less-exalted citizens, the Benedictine order assembled a massive landholding, much of which was vineyard. Cluniac monasteries spread across Europe. The donors were motivated not by generosity but by a desire to atone for an indulgent lifestyle that ran contrary to the church's teaching. Provided the donations were generous enough, being appropriate to the donors' means, the monks would intercede with the Lord on their behalf and grant them a clean slate, allowing them to pursue their less-than-sacred lifestyle with a clear conscience. The monks themselves were not immune to temporal pleasures and in time came to adopt the feasting habits of their benefactors, fuelled by the produce of their vineyards. Monastic asceticism and strict adherence to Benedict's Rule of prayer and moderation gave way to excess and dissipation. The good life took its toll and many monks grew florid of face and full of figure.

In their midst were some who found the dissolute lifestyle repellent, and in 1098 a breakaway group led by Robert de Molesme established a *Novum Monasterium* at Cîteaux, some 10 kilometres east of the Côte d'Or. It was an area of marshy woodland and the name derives from the old French *cistels*, meaning reeds, which grew in abundance there. In time the new Cistercian order took its name from Cîteaux. The land for the new monastery was granted to Robert by the Viscount of Beaune, and other land was granted by the Duke of Burgundy, Eudes I, including a vineyard at Meursault. Fourteen years after its foundation Cîteaux was boosted by the arrival of Bernard de Fontaine, son of the lord of Fontaine, accompanied by a band of thirty followers. He rapidly became the driving force of the new order, instigating a fearsome work ethic that distinguished it from the Benedictines. Where the latter

8 CÔTE D'OR

administered and supervised their vineyards, the Cistercians worked the land themselves. To say they worked themselves to death is hardly an exaggeration – in the early years of the order a Cistercian was unlikely to live past his thirtieth birthday.

The order expanded at an extraordinary pace. Daughter houses sprang up rapidly, including the abbey of Tart, whose cellar and vineyard still exist at Clos de Tart in Morey-Saint-Denis, and the Abbaye de la Bussière, which is now a luxury hotel. By the time of Bernard's death in 1153 about 400 Cistercian monasteries had been established across Europe, and a hundred years later this figure had increased to 2,000. As a consequence they enjoyed massive influence even if they did not wield outright power – much like Google, Amazon, Apple, Facebook, TikTok and their ilk today. In vinous terms, however, the Cistercians' most impressive legacy is the Clos de Vougeot, the remarkable 50-hectare block of vineyard they created over a period of centuries, starting in 1100.

Cistercian monasteries were required to be self-sufficient and wine was a basic necessity, a safe and nourishing drink at a time when a potable water supply wasn't always easy to find. To go with their meals the monks were entitled to a *hemina* of wine per day, about half a pint, and wine was also required for sacramental purposes. But the swampy land at Cîteaux was unsuitable for vines so they moved westwards to Vougeot, where they were granted their first lands in 1100. Other donations soon followed and gradually the roughly rectangular block of vineyard still in existence today took shape. Throughout the twelfth century the Cistercians were also acquiring vineyard land in many other Côte d'Or communes such as Chambolle, Vosne and Volnay. They built a winery in the heart of their Vougeot vineyard and quickly established a reputation as master winemakers. They brought a new rigour to winemaking, studying the land to see which plots yielded the best wines, working intensively and methodically. Because the monks could read and write they could keep records, gradually building a picture of the côte and developing the idea of a *cru* – a defined area of vineyard that yielded wine with an identity of its own, similar to those around it but observably different too. Vineyards that regularly ripened early or late were noted, as were ones that produced stronger or lighter wines, and so on and on, with all observations recorded. In modern parlance they assembled an ever-evolving database that was then used to inform and guide decisions and practices in vineyard and cellar. It was a Herculean task requiring manual labour on a scale that could not be met by their own ranks, so they boosted their numbers by recruiting lay brothers who wore brown habits in contrast to the monks' white.

The Cistercians were made for the côte and it for them; they released its potential and were in turn rewarded with wines of superlative quality. In time, boundaries and divisions between the different *crus* came to be marked formally, often enclosed by the building of a wall to form a *clos*. Then it was time to name them, starting the process whereby the differentiated vineyards, noted for the particular style and quality of their wines, could be easily identified. Thus were born *les climats*, vineyard parcels of unique character, 1,247 of which were granted UNESCO World Heritage status centuries later, in July 2015.

Clos de Vougeot remains the most famous *climat*. It is not clear when it was completely enclosed by walls, perhaps sometime in the 1330s though it may have been later. The eponymous château, as distinct from the vat house and cellar, was built on the orders of the abbot Dom Jean Loisier in 1551. Though not a lavish edifice – the façade is notably austere and bereft of ornamentation or architectural embellishment – it is indicative of the wealth the Cistercians had acquired by this time. They were greatly enriched by donations of land from knights departing for the Holy Land on crusade, keen to buy some insurance with the man above should calamity befall them on foreign fields. As with the Benedictines centuries earlier, the Cistercians had now grown plump and the piety that distinguished them in their earlier years had waned and was no longer practised with such fastidious purpose. From this time on the order was in gradual decline until the time of the Revolution when it was abruptly dispossessed of its remaining, though still extensive, land holdings.

The Cistercians had left their stamp, however. The potable fruits of their labours are long gone but their memory is etched on the landscape in hundreds of demarcated vineyards, the boundaries of which remain largely unchanged today. They established the Côte d'Or's template: the concept of carefully categorized vineyards is the work of the Cistercians. Vougeot is the historic hub of the Côte d'Or. It could be said to be the cradle of modern Burgundy, the starting point for the region and its wines as we know them today.

In the fourteenth and fifteenth centuries the ecclesiastical influence of the monastic orders was matched by the temporal power of the four Valois dukes who ruled over Burgundy from 1363 until 1477. All were possessed of evocative sobriquets: Philippe le Hardi, Jean sans Peur,

10 CÔTE D'OR

Philippe le Bon and Charles le Téméraire: the Bold, the Fearless, the Good and the Reckless. This ducal quartet left a lasting imprint on Burgundy, seen most visibly today in Beaune's Hôtel-Dieu, home to the Hospices de Beaune until 1971. It was built by Nicolas Rolin, chancellor to Philippe le Bon, and construction started in 1443.

Philippe le Hardi earned his moniker for his bravery at the age of fourteen at the battle of Poitiers, and he continued in like spirit in Burgundy after his installation as duke by his father King John II of France. By marrying his predecessor's widow, Margaret of Flanders, he greatly expanded his duchy, which came to resemble an independent kingdom, with him and his successors as monarchs. He is best remembered for his banning of the Gamay grape in 1395, though perhaps he should have stuck to expansion and administration of the duchy, where his efforts met with greater success. His grandson Philippe le Bon was still railing against Gamay's shortcomings and ruling against its use in the côte over half a century later, determined that it should not be allowed to besmirch Burgundy's exalted reputation.

It was during Philippe le Bon's tenure as duke that Burgundy's fortunes reached their apogee. Sitting at the heart of western Europe the region enjoyed tremendous prosperity as a trading hub, particularly on the north–south axis from the North Sea to the Mediterranean. The arts were patronized and craftsmen such as silversmiths and jewellers found a ready market for their products. But it all came to a sorry end when Charles le Téméraire, reputedly fond of the jewellery himself, gave vent to his bellicose ambitions by going to war with his neighbours and was killed in battle in 1477 trying to conquer Lorraine. Burgundy's independent existence was over, and thereafter it was incorporated back into France.

Life didn't always glitter in the age of the Valois dukes, however; recurrent outbreaks of the Black Death saw to that, as it struck high and low with no respect for its victims' station. Yet whatever its depredations it didn't derail the development of the wine business, which saw the emergence of a mercantile middle class in the fourteenth century, many of whom – in Nuits-Saint-Georges and Beaune, for instance – owned small plots of vineyard that they tended as a sideline to their principal occupation. By the end of the fifteenth century a quarter of Dijon's workforce was made up of smallholder wine growers. Substantial plots were divided and sub-divided, resulting in the emergence of new vineyards quite separate from those of the aristocracy and church. This further embellished the mosaic of *climats* and *lieux-dits* that today form the substance of the Côte d'Or. Notwithstanding this development, the twin pillars of aristocracy and church remained the two defining forces in Burgundy for another couple of centuries, up to the time of Revolution.

The monasteries' golden age was over, though, and steady decline was to be their lot in this period as they sold off or rented out prized vineyards to boost diminishing coffers. The city of Dijon was growing in prosperity, with many wealthy citizens in search of trophy assets to boost their standing. The church had assets aplenty and they found ready buyers: La Romanée was sold in 1631 and Clos de Bèze twenty years later. About a century after that, in 1760, the Prince de Conti purchased the most illustrious vineyard of them all, to which his name has been appended ever since: Romanée-Conti. Perhaps luckily for him he died in 1776, some years before developments outside the Côte d'Or had a shattering influence on it, shredding the ownership model that could be traced back to the founding of Cluny and Cîteaux.

In the simplest terms the French Revolution was caused by bankruptcy and starvation: the state was bankrupt because of involvement in the American war of independence and the people were starving because of a series of disastrous harvests. Burgundy itself was not gripped by the same revolutionary fervour as were the cities, and the privations that caused the Revolution may have had a greater impact elsewhere, but the Côte d'Or felt the full force of its consequences. It released a flood of pent-up anger directed at the aristocracy and the church, and in time it turned on itself and some of its initial leaders followed the noblemen to the guillotine. France finished the eighteenth century in a maelstrom of violent upheaval.

Out of this emerged one of the most compelling figures in world history, Napoleon Bonaparte, who rose to prominence through a series of brilliant military victories. In the Côte d'Or at the same time the ownership map was being redrawn with equally compelling force. The côte may have been spared excessive bloodletting but the cash-strapped state was not blind to the value of the vineyards, which were promptly seized and sold off as *biens nationaux* or national assets. Clos de Vougeot, along with its château and other buildings containing four giant presses, was sold at auction in 1791, though the highest bidder was unable to pay, so it was left in the care of the Cistercian cellarmaster Dom Goblet. He distinguished himself in the service of his new secular masters, who rewarded him handsomely for his efforts. Some years later it ended up in the hands of Julien-Jules Ouvrard, son of Napoleon's banker, and it remained in his family's possession until, in 1889, a century after the Revolution, it was sold to a group of négociants. The process of fragmentation, which today sees the *clos* with some eighty owners, had begun. Its fate was repeated up and down the côte, though usually more rapidly than this. The concept of distinct *crus*, many of them enclosed by a wall to create a *clos*, whose character was defined by the particular attributes of each, was irreversibly changed to the point where knowledge of who made the wine is now more important than the name of the vineyard.

Where the Revolution had an immediate impact on the côte, Napoleon's influence, in the shape of the *Code Napoléon* dictating that an inheritance should be divided equally between all offspring, was more long term and still has a significant influence on the ownership of vineyards and domaines. The Revolution led to fragmentation while the code led to what can only be described as pixelation with vineyards minutely subdivided, to the point where a treasured few rows in a top *grand cru* may yield less than a standard barrel of wine per year, necessitating the construction of a barrel that is custom made to fit the wine.

The trauma of the Revolution followed by the Napoléonic wars, a quarter-century of turmoil and conflict in which France lost 1.5 million men, got the nineteenth century off to a tormented start, but thereafter things settled and the business of making wine became more structured. In the decades leading up to and past the middle of the century the Côte d'Or started to warrant its designation as the 'Golden Slope'. In 1790 its name had been given to the new *département*, which took in much more than the vine-clad slope, by André-Remy Arnoult, a deputy in Dijon. Whether 'Or' means gold at all or not is the subject of much debate; a more prosaic conjecture is that it is an abbreviation of 'Orient', referring to the côte's easterly alignment.

It also began to be written about and codified. In 1831 Denis-Blaise Morelot set out the first carefully documented classification of the côte's vineyards in his paper *Statistique de la vigne dans le département de la Côte d'Or*. This work was succeeded by Dr Jules Lavalle's more comprehensive *Histoire et statistique de la vignes des grands vins de la Côte d'Or* in 1855, the same year as Bordeaux's famed classification. Included in it were detailed maps and a system of vineyard ranking, the precursor to that formally adopted in the 1930s. Compared to Bordeaux's classification, Lavalle's is largely forgotten by the general public yet it was no less significant in its time.

The railway came to Dijon in 1851, opening up the Paris market and transforming Burgundy into the capital's vineyard. The Hospices de Beaune auction was first held in 1859 and has become the largest charity wine auction in the world. These were good times for the côte though decline and calamity lay ahead. In time, the railway extended further south to the Languedoc, from where it brought back vast quantities of cheap wine that undercut Burgundy. Perhaps travelling with it was something much more destructive than mere competition: phylloxera.

The deadly aphid was first noted in Provence in 1863 and had reached Meursault by 1878. It caused vines to die by eating their root systems and soon wrought devastation across the vineyards of France. An industry and a way of life that were stitched into the nation's fabric were threatened. Desperation prompted remedies that seem comical in retrospect but these were people fighting for their livelihood; anything that promised even a faint hope of success was to be tried. It is hard to stifle a laugh at the suggestion that a live toad buried under each vine was the solution. One treatment that did work was the injection of carbon bisulphide into the soil, which was deadly to phylloxera and to much else besides, including the vine if applied too liberally. It was a tedious and laborious process, however, and was effective only because it dealt a sledgehammer blow with, in today's parlance, an unacceptable level of collateral damage, including sometimes the workers applying the noxious treatment when it caught fire. Eventually, the solution of grafting onto American rootstocks was discovered, a practice adhered to still.

A small positive was that phylloxera gave impetus to a more rigorous and scientific examination of viticulture and oenology, leading to a greater understanding of both and the establishment of a school of viticulture – the 'Viti' – in Beaune in 1884, as well as the *Station Oenologique* in 1900 on the *péripherique* road, where the Bureau Interprofessionnel des Vins de Bourgogne is now housed. Nonetheless, phylloxera cast a long, baleful shadow and when allied to the twin scourges of oidium and mildew, it ushered in a near-century of difficulty, adversity and hardship. This was compounded by war and economic turmoil, leaving the Côte d'Or in a sorry state by the 1950s, from which it only recovered by the close of the twentieth century.

It would be difficult to overstate the destructive legacy of phylloxera. The changes wrought by the Revolution, hugely significant in

14 CÔTE D'OR

themselves, are matched if not exceeded by phylloxera. It's not easy to say which had the greater or more lasting effect. The Côte d'Or was indelibly stamped and shaped by these two forces, the one overt and violent, the other hidden, though no less destructive. The monks left an imprint on the Côte d'Or, overlaid now by those malign forces, which initiated a degree of change over a century not seen in the previous millennium. The French Revolution and phylloxera are the violent and destructive parents of today's Côte d'Or.

THE TWENTIETH CENTURY

By 1900 the buoyant days of the 1850s must have seemed very distant, even for those with a memory of them. The face of the côte was changing rapidly as the vineyards were grubbed up and replanted with vines grafted onto American rootstocks. For centuries prior to this vineyards were renewed by *provignage*. Cultivation was completely manual and the ground under the vignerons' feet was composed of decaying roots, old vines and soil, rich in life. The post-phylloxera planting was done in regimented rows, which meant that horses could be used for ploughing. The equine age was short lived, however, and was brought to an end by the large-scale adoption of vine-straddling tractors that look like mechanical giraffes – *tracteurs enjambeurs* – after the Second World War. The horses' reappearance at some prestigious domaines in the early years of the twenty-first century elicits nostalgic nods of approval from first-time visitors, mistakenly assuming that this is a return to ancient practice.

An indirect consequence of phylloxera and the other afflictions that ravaged the côte and all France in the closing decades of the nineteenth century was the rise in fraudulent wine production, a development echoed twenty years later in the United States when Prohibition gave rise to bootleg whiskey. As wine production plummeted, every manner of shady character stepped forward to make good the deficit. Wine was made from raisins, wine was made from sugar beet – indeed, such was the scarcity of genuine wine that these fraudulent wines were openly produced in huge quantity to service demand.

It had never been definitively laid down just what constituted 'Volnay', 'Beaune' or 'Chambertin'. Names such as these had strong images and functioned more as brands than as indications of geographical origin. As such, they were ripe for imitation and adulteration. It wasn't long before the authorities realized that legislation was needed to protect consumers, the first of which was passed in 1905. Thus began the three-decade process that resulted in the *Appellations d'Origine Contrôlée* regulations of 1935, though much else intervened before that, most brutally the First World War. The fighting never reached the côte but the young men were torn out of it, along with the rest of France, as attested by the memorials inscribed with the names of many well-known winemaking families in every village.

In 1919 more exacting legislation was passed to establish provenance as a wine's defining characteristic, by allowing local authorities to designate boundaries for the various communes. Prior to this a wine made in Morey, for instance, might be sold as the better-known Gevrey-Chambertin. Such restriction was not to the liking of the large négociants who controlled the trade at the time, and it set them on a collision course with the growers who cared about a wine's origin, while the négociants bought and blended wine from here and there and then labelled it loosely, perhaps as 'Beaune'. Two growers took up the cudgels in the fight for authenticity: Henri Gouges of Nuits-Saint-Georges and the Marquis d'Angerville of Volnay.

The zeitgeist was in the growers' favour, though the day was not won without a fight as the négociants sought to preserve their *modus operandi*, which might see them rustling up a wine to order; the restrictions, controls and regulations the growers were pushing for would cramp that creative blending style. Gouges and d'Angerville were among the first to use labels as a guarantee of authenticity, which was a step towards the appellation controlée regulations introduced shortly afterwards. That a wine should be what it claimed to be became Gouges' life's work. In 1930 Pinot Noir was specified as the côte's noble red grape and in 1935 the nationwide appellation controlée regulations came into place, specifying details such as ripeness and yields, as well as cultivation and winemaking methods, criteria that the wine had to meet if it was to be granted appellation status. There have been various tweaks since, such as the upgrading to *grand cru* of Clos des Lambrays in 1981 and La Grande Rue in 1992.

The feisty growers were not popular with the négociants who stopped buying wine from them, shutting that route to market – what market there was – and thus prompting the move to domaine bottling. These were not happy times in the Côte d'Or, the misery being compounded in the early 1930s by Prohibition in the United States and worldwide

16 CÔTE D'OR

economic depression. Having wine to sell – a blessing today – was little better than a curse then. A home-grown solution of sorts was the founding in 1934 of the *Confrérie des Chevaliers du Tastevin*, a bibulous fraternity of burgundy lovers, the reasoning being that if the wines could not be sold then they might as well be drunk and enjoyed in convivial circumstances. In grim times, wine's gentle narcotic glow blurred cares and soothed spirits.

There was cause for cautious optimism towards the end of the decade. Prohibition had been lifted and a chapter of the Chevaliers was established in New York in 1939. And then war came again. Hitler's panzers smashed through the Ardennes in May 1940 and although the death toll wasn't as high as in the previous war the swastika soon flew over the côte. The occupiers' rapacious demands for wine were met with every method of evasion, though so many cellars were reportedly bricked up to hide the best wines that it is a wonder there were enough bricks for the task.

In the aftermath of the Second World War the European wine business was in a sorry state, riven by war and buffeted by economic hardship. In the Côte d'Or the ravages of phylloxera were still a vivid memory. Owners of celebrated vineyards fought tenaciously to retain their precious old vines – the ungrafted vines in Romanée-Conti were only grubbed up in 1945, before replanting two years later. Nearly a century of affliction had left Burgundy's heartland battered and cast down. From the prosperous vantage point of the twenty-first century it is difficult to grasp the sorry state of the wine world some seventy years ago. The fiscal frenzy that grips today's côte, where astronomical sums are paid for slivers of favoured vineyard, could never have been foreseen.

In such circumstances it is understandable that the vignerons, who only knew winemaking as a grim and largely profitless business, should grasp at any remedy that promised abundance and prosperity. Thus began the era of fertilizers and weedkillers, used like magic wands, with the promise of short-term gains obliterating any consideration for the long-term consequences. It was a prescription for plummeting quality in the race for quantity yet who, in the same circumstances and after generations of decline, would not have done the same? For a period the land was anything but valued; it wasn't nurtured and cherished and prized like it is today, it was there to be exploited, flogged with every manner of chemical stimulant and remedy. It was a thoroughbred being treated like a dray horse and soon it began to behave as such, yielding vapid wines of little character and no excitement. Many of these were then discreetly stiffened and given some spine with the addition of something hearty from the Midi or Algeria. The fertilizers also interfered with the balance of the soil and altered the wines' acidity, a deficit that had to be rectified in the winery.

Dismal vintages, such as are not seen today, were regular occurrences. The hundredth anniversary of the discovery of phylloxera in France, 1963, was appropriately awful. It had been a brutally difficult century for the Côte d'Or and it was not an anniversary to be celebrated. The 1970s, though not fondly remembered now, saw some moves in the right direction, with domaine bottling increasing, adulteration becoming less common thanks to stricter EEC/EU regulations and an increasing awareness that trying to make a hearty wine from Pinot Noir was a fool's task. Gradually, too, the depredations of phylloxera began to fade. Heretical as it may sound there is an argument for phylloxera being ultimately a force, perhaps not for good, but certainly for improvement in viticulture and ultimately wine quality. Recovery from its lingering after effects was snail-slow, but by the 1980s the Côte d'Or was ripe for renewal.

2
VINEYARDS AND VITICULTURE

Terroir. If the complex and storied Côte d'Or can be summed up in one word it is 'terroir'. It's a word that is bandied about, sometimes positively to give a clear explanation of a wine's character, at other times negatively to create a smokescreen to hide a wine's shortcomings. But what is 'terr-wahr'? It defies easy definition and resists translation. 'Place' is probably the best equivalent word in English, used in the broadest sense and encompassing a sense of place as well as an exact physical location. Terroir includes the soil and subsoil, the climate and microclimate, the elevation and aspect, the inclination and drainage of a plot of vines, and it even extends to the people who tend the vineyards and make the wine. Without the hand of man terroir sits mute; how the winemaker chooses to exploit it adds another dimension and constitutes an essential component of the whole. In short, while terroir refers primarily to the soil and what lies beneath, it also encompasses everything about a particular place that has an influence on the wine made there and which gives it a character and identity that is not replicable elsewhere. A similar wine may be made a stone's throw away but because of a small shift, perhaps in soil or slope, there will be subtle, and in some cases dramatic, differences between them.

The way a vineyard is tended is crucial to harnessing and releasing its potential, and therefore the influence exerted by the vignerons is central to any understanding of terroir. Yet including them in a broad definition of terroir can raise eyebrows; after all, people come and go while the place itself is fixed. True, but over the centuries the Côte d'Or has been stamped by the people, measured and marked by them, tended and tamed by them, and their influence cannot be ignored. Today, that influence is actively benign, people working in harmony with place, yet in the decades following the Second World War the people slipped out of step with their place and their influence became mistakenly malign. The land was exploited in a rapacious sense rather than cherished as the vignerons' prime asset.

> ## A terroir tale
>
> Saturday 2 March 2024. The tasting was tight-focused, consisting of seven vintages of Domaine Dujac, Gevrey-Chambertin *premier cru* Aux Combottes: 1994, 1999, 2001, 2002, 2003, 2005 and 2009. Star vintages on the night were the 2002 and 2009, though all were agreed that the 2005 might shine brightest with another decade's age. Aux Combottes lies at the southern extreme of the commune and is surrounded by *grands crus*: to the north, Latricières-Chambertin; to the east, Mazoyères-Chambertin; to the south, Clos de la Roche; and to the west, the Monts Luisants *lieu-dit* of Clos de la Roche. All of which prompted the question: Why had *grand cru* status eluded Combottes? The answer came by way of two more Dujac wines, both 1999 *grands crus*, Clos de la Roche and Echézeaux. They were effortlessly better than the Combottes, especially the gracious and sublime Echézeaux. Terroir spoke in clear and convincing fashion that evening.

Today, the terroir gospel is preached fervently, perhaps a little too fervently at times. There is hardly another topic in the whole world of wine that provokes such passionate debate and proponents on either side of the argument have well-stocked arsenals to back up their belief that it is either a God-given set of beneficial characteristics that only a handful of favoured vineyards possess, or little more than a load of old hokum. The concept of terroir may now be at the point where it is too sanctified, accorded too much unquestioning belief, preached with too much zeal. When propounded as dogma it comes across as a near-mystical set of beliefs that cannot be questioned or rigorously examined: you are either a believer or you are not seems to be the message. It does exist, however; it is not simply a collection of hazy notions spun into something more substantial. The downside is that winemakers can use it as a smokescreen to explain flaws or off flavours, though less frequently than

20 CÔTE D'OR

heretofore, and sommeliers have been known to invoke its mysteries to excuse an obviously faulty wine.

PRINCIPAL VINEYARDS – CÔTE DE NUITS

For many people, their mind's eye picture of the Côte d'Or stretches from Gevrey-Chambertin, the first commune that is home to *grand cru* vineyards, to the last, Chassagne-Montrachet. Until recently this was a safe mental attenuation, lopping off the northern and southern country cousins and not paying much heed either to some others in between, such as Prémeaux-Prissey or Saint-Romain. Not a lot was lost in the process and memory space could be reserved for the wines that really mattered. Such an exercise today would be ludicrous, ruling out a host of yet to be celebrated vineyards at the northern and southern extremities of the côte as well as others in between.

A brief overview of the vineyards, running from north to south, begins in the outskirts of Dijon, whose urban sprawl has engulfed land that was previously home to the vine. The first vineyards are in Chenôve, though it is at Marsannay-la-Côte that the shopping centres and light-industrial zones segue into unbroken vineyard. The slopes are gentle here and, as yet, there are no *premiers crus*, a situation that may change in the future if current efforts to get a proportion of vineyards upgraded are successful. Some producers already use *lieu-dit* names such as Longeroies, probably Marsannay's best site, Clos du Roy, which lies in the Chenôve commune, and Es Chézots, which is noted as much for how it should be spelt (Les Echézeaux, Echézots) as for the quality of its wine. 'Snail's pace' doesn't begin to describe the rate of progress in gaining *premier cru* status for the best vineyards, but a tasting of Sylvain Pataille's best wines makes the case more compellingly than the reams of paperwork required.

Continuing south through Couchey, which is included in the Marsannay appellation, we come to Fixin, whose handful of *premiers crus* are the highest in the commune, all lying above 300 metres and abutting the forest. In total they amount to about 20 hectares and the remaining 100-plus hectares qualify for the Fixin or Côte de Nuits-Villages appellations. The best known of the *premiers crus* is Les Hervelets, a *climat* that includes the *lieux-dits* of Le Meix-Bas and, confusingly, Les

Arvelets. The latter may be made as a separate wine but this is seldom done; Hervelets is the name to look for.

Sandwiched between Fixin and Gevrey-Chambertin is Brochon, whose band of southerly vineyards is included in the Gevrey appellation. The best known are Les Evocelles and Les Jeunes Rois, the former high on the slope and once easily spotted thanks to the Domaine de la Vougeraie section being planted *en foule*, meaning in a crowd, at a density of 30,000 vines per hectare. They have abandoned that experiment, since the authorities now insist on a minimum spacing of half-a-metre between vines. Students of orthography will note that the corner of Evocelles that crosses the commune boundary into Gevrey changes its spelling to Evosselles; others will scratch their heads in bafflement.

The paucity of highly ranked vineyards encountered thus far is amply rectified in Gevrey-Chambertin, home to nine *grands crus* and a slew of *premiers crus*. In each category there are vineyards that fully justify their status, none more so than Chambertin and Chambertin-Clos de Bèze, a pair of the Côte d'Or's most esteemed vineyards. The latter may be labelled simply as 'Chambertin' but the reverse is not allowed. At their best these neighbours yield wines of majesty and substance, capable of long ageing, the Chambertin perhaps sturdier and stronger than the slightly lighter footed Clos de Bèze. Together they form an oblong block of some 28 hectares, about 300 metres wide and less than a kilometre long. The *Route des Grands Crus* forms their eastern boundary and travelling along its north–south axis the slope is barely perceptible; a walk up towards the forest and back is needed to notice the roughly 25-metre rise from bottom to top.

> ## Burgundy or Bourgogne?
> The powers that be have decreed that henceforth Burgundy will be known as Bourgogne, explaining their reasoning thus: 'Bourgogne is the only wine-producing region in France whose name is translated into different languages: "Burgundy" for English speakers, "Burgund" for Germans, "Borgogna" in Italian, to name but a few. However, the word "Bourgogne" is on every label … That's the reason why we decided to stop translating the word "Bourgogne", whatever the country or language. The aim is to help consumers find their way by ensuring coherence between our wine labels and the name of the region where the wines were created.' Well, good luck with that. 'Bourgogne' rolls easily off the tongue but I don't ever see it replacing the sonorous, suggestive ring of 'Burgundy', at least not in the English-speaking world.

22 CÔTE D'OR

The seven other *grands crus* are Chambertin satellites and all appropriate its exalted name to gain recognition by way of reflected glory, as with Montrachet in the Côte de Beaune, though the Chambertin 'clan' is more scattered and numerous. They claim 'Chambertin' by virtue of being contiguous with it or Clos de Bèze, though Ruchottes' connection is fingertip slim and calls to mind Michelangelo's Creation of Adam. The seven are Chapelle-Chambertin, Charmes-Chambertin, Griotte-Chambertin, Latricières-Chambertin, Mazis-Chambertin, Mazoyères-Chambertin (usually labelled as Charmes) and Ruchottes-Chambertin. Mazis borders Clos de Bèze to the north with Ruchottes above it, reaching up to the tree line above 300 metres, while Latricières is Chambertin's southern neighbour. The western flank of all these vineyards, with the exception of Mazis and part of Clos de Bèze, is cheek by jowl with the forest, meaning that the vines there go into the shade of the trees much earlier in the day than those to the east. As such, the siting of these *grands crus* doesn't accord with the oft-repeated tenet that they lie in mid-slope, cushioned above and below by lesser *crus*. The remaining quartet – Chapelle, Charmes, Griotte and Mazoyères – lie on the other side of the *route* and in the case of the latter reach right down to the D974 main road, where there is virtually no slope, a hardly ideal situation that risks devaluing the Chambertin name. The wines from the satellite seven can be excellent even if they never surpass the heights achieved by the first pair.

Some two-dozen *premiers crus* cover over 80 hectares and include at least one – Clos Saint-Jacques – that is worthy of *grand cru* status. It sits above the village with a perfect south-east exposure, plumb in the centre of a crescent of *premiers crus* that girds the hillside. So obvious is its *de facto grand cru* standing that nobody bothers to agitate for its elevation. Supposedly, it was overlooked when the *grand cru* gongs were being handed out because it was not contiguous with Chambertin, though a more colourful suggestion blames a cussed previous owner who so irritated the authorities that they were never going to confer top-rank status on his vineyard. Clos Saint-Jacques' immediate neighbours on either side, Lavaut Saint-Jacques and Les Cazetiers, are not quite so renowned, though both sit at the top of the *premier cru* tree.

The Côte de Nuits stretches to its widest at Gevrey. From its western extreme at the pinpoint of La Bossière it is over 4 kilometres across to the 18-hectare La Justice vineyard which is located on the 'wrong' side of the D974. Though flat, it can produce vigorous wines well worthy of their appellation.

After the glamour of the Chambertin name it is understandable that Morey-Saint-Denis carries less cachet, less immediate recognition. It is a compact commune, not 2 kilometres from north to south, and is home to four-and-a-sliver *grands crus*, the sliver being Bonnes Mares, which is generally treated as if it resided wholly in next-door Chambolle-Musigny. The four divide easily into two pairs: Clos de la Roche and Clos Saint Denis to the north, and Clos des Lambrays and Clos de Tart to the south.

Unlike Gevrey's *grands crus* these do sit at mid-slope and straddle the commune in linear succession. Though all four are '*clos*', it is Clos de Tart that does justice to that designation, being enclosed by walls in a fashion that is largely absent in, for instance, Clos de la Roche where you can park your car besides the *Route des Grands Crus* and stroll into the vineyard. Until recent improvements at Lambrays and Tart the northern pair were Morey's standard bearers, with Roche generally regarded as the better of the two, though its greater consumer visibility is down to its size – at a shade under 17 hectares it is nearly three times the size of Clos Saint-Denis. Between them they encompass a dozen *lieux-dits*, including the evocatively named Maison Brûlée that abuts the village dwellings. The name probably derives from the sacking of the region in 1636 by Austrian troops of the Emperor Ferdinand II, with whom France was at war.

While Clos des Lambrays and Clos de Tart are roughly equal in size (8.8 and 7.5 hectares respectively) their shapes differ markedly, the latter's neat rectangle making the former's boundaries look ragged by comparison. In the past it didn't help Morey's standing that this pair seldom lived up to their potential, a situation that no longer pertains thanks to recent massive investment, with concomitant surge in quality, at both domaines. (For more on Clos des Lambrays and Clos de Tart, see Chapter 4.) Morey's *premiers crus* cluster mainly downslope of the *grands crus* though some of the best such as Monts Luisants lie above, between 300 and 350 metres. It is best known as a *premier cru* for red wine and also for Domaine Ponsot's famed Aligoté, proof that marvellous wine can be made from this previously overlooked, though now resurgent, grape. At *village* level Clos Solon, adjacent to the D974, yields a memorable wine in the hands of Jean-Marie Fourrier. Look out also for En la Rue de Vergey, the memorably named though seldom seen neighbour of Clos de Tart, which punches above its *village* status in the hands of Bruno Clair.

24 CÔTE D'OR

Chambolle-Musigny is noted for wines of grace and elegance – with more substance in recent years – yet its pair of *grands crus* can hardly be considered as two sides of the same coin; they are more differentiated than that. Bonnes Mares and Musigny are the opposite poles of Chambolle; in style the former looks north to the commanding structure of Gevrey-Chambertin, while the latter casts its glance south towards the grace and finesse of Vosne-Romanée. This is not to suggest that they don't each have a clear identity, more to illustrate the difference between them. The pair are separated by the village itself and a swathe of *premiers crus* that runs between them. Travelling from Morey, Bonnes Mares is the first vineyard you encounter, a substantial rectangle of 15 hectares that crosses the commune boundary, with about 90 per cent of it in Chambolle. It is difficult to generalize about Bonnes Mares because there is a radical difference between the soils in the upper and lower sections of the vineyard. What can be asserted is that by comparison with Musigny it produces a heartier wine, with more spice and something of a *sauvage* character. If it lacks something of the perfumed grace of Musigny its impact is more immediate; visceral to Musigny's sensual.

Climat or *lieu-dit*?

Understanding the difference between a *climat* (klee-mah) and a *lieu-dit* (lyugh-dee) and being able to explain it without confusion is a Sisyphean task. In truth they can almost be regarded as synonymous and are often used interchangeably; the difference lies mainly in their derivation. The use of both terms dates back centuries and a tentative distinction suggests that *climat* is the term favoured by *vignerons* to refer to a clearly delineated parcel of vineyard with its own distinct terroir, while *lieu-dit* is favoured by cartographers. Also, *climat* usually refers to *premier* and *grand cru* vineyards, though not always. Thus *climat* might be considered the senior term. The parcels they refer to on the map are frequently identical, sometimes overlap and sometimes do not.

An example will illustrate the confusion though hardly clarify it. In Morey-Saint-Denis the *grand cru climat* Clos de la Roche encompasses eight *lieux-dits*, including Monts Luisants. However, only the lower section (some 4 hectares) of the Monts Luisants vineyard, which covers a total of about 11 hectares, lies within Clos de la Roche. The middle section is *premier cru* and is both *climat* and *lieu-dit*, and the upper section, close to the tree line, is *village* and only *lieu-dit*. Wine from any section could have 'Monts Luisants' in some shape or form on the label, though only the *premier cru* is virtually certain to have.

Musigny overlaps the top corner of Clos de Vougeot and comprises three *lieux-dits*: Les Musigny, Les Petits Musigny and La Combe d'Orveau. Its eastern boundary is completely formed by the *Route des Grands Crus* – or so it appears until a close examination of the map reveals a shred of vineyard that lies across the road from the main body of the vineyard. It sits on a step of ground at the top of Bertagna's *monopole* Clos de la Perrière and is home to a couple of hundred individually staked vines. It is so small that it is hardly worthy of mention but because it belongs to one of the côte's most celebrated of all *grands crus* it is worth cultivating. Pascal Marchand of Marchand-Tawse vinifies a magnificent wine from this tiny origin. A final quirk that distinguishes Musigny from all other Côte de Nuits *grands crus* is that it is permitted to plant Chardonnay there, though de Vogüé is the only producer to make a Musigny *blanc*.

It's a hackneyed assertion that Musigny is the Côte d'Or's queen while Chambertin is the king, a memorable, if hardly profound, observation that might be dismissed as an old nugget of *faux* wisdom. Yet it stands up to scrutiny. The power and concentration of Chambertin is absent in Musigny, replaced by more moderate qualities of elegance and poise. There is strength, but it is the finessed strength of the ballet dancer not the overt weightlifter's version. Its qualities have been the cause of much superlative frenzy over the centuries thanks to the extraordinary intensity of complex scents and unfolding, layered fruit flavours. A significant portion of the vineyard is owned by de Vogüé, though perhaps the most sought-after is Roumier's, a circumstance brought about by a combination of superb quality and tiny quantity.

Of Chambolle's *premiers crus* Les Amoureuses, downslope from Musigny, stands apart and is accorded putative *grand cru* status, much like Clos Saint-Jacques in Gevrey. Thanks to quarrying in previous times, Amoureuses presents a more jumbled appearance than its neighbours and the derivation of its name is fertile ground for speculation. Perhaps it was a venue for torrid trysts; more prosaically it is suggested that the soil when wet clings to footwear with a lover's grip. All who walk the vineyard are struck by the contrast between its fractured and rock-strewn terrain and the remarkable grace of the wine it produces.

Though Musigny and Clos de Vougeot share the same vineyard classification and indeed share a boundary for a couple of hundred metres, along which they are separated by a literal stone's throw, a huge gulf in renown divides them. Where superlatives rule the roost with Musigny

26 CÔTE D'OR

it is hard to write about Clos de Vougeot without slipping into cliché, trotting out the rote statistics used for generations to illustrate its short-comings. It is a roughly square, 50-hectare block of vineyard, a little longer on the diagonal that runs from the south-east corner up past the château to Musigny. The Côte d'Or's usual clutter of tiny, variously shaped vineyards, threaded with roads, tracks and dry stone walls, is absent here, where the vines seem to stretch to the horizon. If it was in Bordeaux it would have one owner and produce two, perhaps three wines; here it has more than eighty, many of whom lay claim to sliv-ers of land so thin that in places the ownership map looks like a bar-code. It did once have a single owner – the Cistercian order – but the Revolution saw them dispossessed and their flagship vineyard sold off as a *bien national*. Fragmentation was slow at first but accelerated through the twentieth century to the point where today's ownership mosaic is a cartographer's delight, or not. The best wines rank with the best of the Côte d'Or, carrying the conviction and energy that should be present in a *grand cru*, but they also serve to highlight the deficiencies of the oth-ers. In some respects the *clos* is the Côte d'Or in microcosm; knowing where the vines lie is useful but who farms them and makes the wine is critical, here more so than in any other *grand cru*, save for Vougeot's southern cousin, Corton, another behemoth that would be improved by some trimming. The recent good news is that socks are increasingly being pulled up and the stereotype outlined above, while still largely valid, is starting to show its age. Clos de Vougeot, the slumbering giant, is beginning to stir.

Clos de Vougeot contains sixteen *lieux-dits* that are not officially rec-ognized and so are seldom seen, apart from Le Grand Maupertuis, used by Anne Gros, and the clever use of Musigni by Gros Frère et Soeur for their wine from the north-west corner of the *clos*, adjacent to Musigny. It seems surprising that almost no other producers use them to create a semi-separate identity although it is doubtful if adding names such as Quartier des Marei Haut or Montiotes Basses would add lustre to the Vougeot name – probably the reverse. At *premier cru* level Vougeot con-tinues to confound, for the most prestigious, in this red-wine heartland, is the white Le Clos Blanc, a *monopole* of Domaine de la Vougeraie.

An oft-cited criticism of Clos de Vougeot is that it runs right down to the main road, with a negligible slope in its lower section, while above it and better sited lie the two *grands crus* of Flagey-Echézeaux: Les Grands Echézeaux and Echézeaux. To all intents and purposes they are considered part of the next commune, Vosne-Romanée, home to the most celebrated vineyards in the world: La Romanée-Conti, La Tâche, La Romanée, La Grande Rue, Richebourg and Romanée Saint-Vivant. Taken together this half-dozen amount to about 28 hectares, not much more than half the area of Clos de Vougeot

It is not possible to overstate the renown in which these *grands crus* are held, particularly the two *monopoles* owned by Domaine de la Romanée-Conti. The eponymous vineyard is a rough square of 1.8 hectares and is marked by a gaunt cross, making it easy to find as you travel up from the village on the small road that runs through Romanée Saint-Vivant. It is hardly an exaggeration to say that it is a place of pilgrimage for wine lovers from across the globe and, conveniently, there is space for a few cars to park next to the vineyard, with clear sight of the sign on the low surrounding wall asking visitors not to walk through it, a request heeded by some: 'Many people come to visit this site and we understand. We ask you nevertheless to remain on the road and request that under no condition you enter the vineyard.' At about six hectares La Tâche is considerably larger than Romanée-Conti and doesn't attract quite the same level of reverence, relatively speaking. La Grande Rue forms a narrow strip between these two and is a *monopole* of Domaine Lamarche, being elevated to *grand cru* from the 1991 vintage. The fourth and smallest *monopole*, at less than one hectare, is La Romanée. Owned by Domaine du Comte Liger-Belair this jewel is contiguous with Romanée-Conti, lying to the west of it on slightly higher ground, and there has been speculation that the two once formed a single vineyard. There is no wall or other marker between them but the relatively unusual north–south orientation of the rows in La Romanée makes it easy to spot the boundary. (For more on these *monopoles* see the respective domaine profiles in Chapter 4.)

Of the remaining *grands crus,* Richebourg and Romanée Saint-Vivant vie for top spot in the affections of burgundy lovers. Richebourg is a wine of flesh and substance, structure and depth, variously described as 'sumptuous', 'opulent' and 'voluptuous', qualities reflected in the plangent ring of its name. There's ballast in Richebourg. It neighbours Romanée-Conti to the north, and the *lieu-dit* at its northern end, Les Verroilles, turns slightly north of east, causing the grapes to ripen a little later than the rest of the vineyard. Domaine de la Romanée-Conti is the principal owner in Richebourg; of the others Thibault Liger-Belair's example is a particular favourite. Romanée Saint-Vivant lies below

28 CÔTE D'OR

Richebourg, close to the village, and takes its name from the nearby abbey of Saint-Vivant at Curtil-Vergy, the remains of which have recently been secured against further decline. The wine is scented, graceful and elegant, and can hardly be matched for delicate intensity, a violin to Richebourg's cello.

Les Grands Echézeaux and Echézeaux don't enjoy the same renown, which is hardly surprising in the case of the latter, given that it includes eleven *lieux-dits* comprising a cumbersome 38 hectares, divided between 59 owners. The name 'Echézeaux' acts as an umbrella under which shelter varied styles of wine. Much the same criticisms that are levelled at Clos de Vougeot apply here – the paramount consideration when searching for quality must be the name of the producer. It is a dictum that applies everywhere in the Côte d'Or, but with heavy emphasis in places like this. Grands Echézeaux, on a barely perceptible slope, is separated from Clos de Vougeot by a narrow road and, with deeper soil delivering more weight in the wine, is generally considered superior to Echézeaux. As a distinguishing mark the 'Grands' is fully deserved.

While the Vosne *grands crus* hog the limelight, two favourite *premiers crus* bookend them: Les Suchots to the north and Aux Malconsorts to the south. They live somewhat in the shadow of their 'big brothers' but superlative wines emanate from both.

After the surfeit of *grands crus* in Vosne-Romanée the next commune south, Nuits-Saint-Georges, is home to none and must settle for the distinction of lending its name to the Côte de Nuits. Because of its size and memorable name it is probably as well known as Vosne, if not nearly as highly regarded. The town in turn takes its name from its most prestigious vineyard Les Saints-Georges, at the southern limit of the commune and reputedly the first plot to be planted in Nuits, in 1000. Efforts to get it upgraded to *grand cru* are ongoing. It is probably the only one of Nuits' *premiers* to warrant promotion, though a case could be made for Aux Boudots right at the other end of the commune, abutting Vosne. A large cohort of *premiers crus* lies in two bands on the hillside, split by the village, which follows the route of the Meuzin river. The Nuits appellation continues south into Prémeaux-Prissey, home to the large *monopoles* Clos de l'Arlot and Clos de la Maréchale.

Thereafter the côte is pinched narrow by rock at Comblanchien and Corgoloin, where vineyards give way to the quarries that form the stony sinew connecting the Côte de Nuits with the Côte de Beaune. The final vineyard contains a little flourish in the shape of Domaine d'Ardhuy, whose impressive building is set back from the road and surrounded by the vines of its *monopole* Clos des Langres.

Côte de Nuits-Villages

This appellation is made up principally of vineyards in Comblanchien, Corgoloin, Prémeaux-Prissey and Brochon, amounting to about 200 hectares, along with a further 100 hectares or so in Fixin that may be labelled as Côte de Nuits-Villages or Fixin. Comblanchien and Corgoloin form the heart of the appellation and red wine makes up the majority of production. Lush vineyard vistas are absent here – quarried rock and finished slabs of stone are the most visible harvest. From a ripe year the wines can offer easy satisfaction and early drinking while waiting for the grander ones to mature.

PRINCIPAL VINEYARDS – CÔTE DE BEAUNE

Travelling south, the unmissable bulk of the Hill of Corton announces the beginning of the Côte de Beaune. The hill is home to an absurdly large band of *grands crus* spread across three communes: Ladoix-Serrigny, Aloxe-Corton and Pernand-Vergelesses. It wraps around the hill, facing east-south-east in Ladoix, continuing to south-east in Aloxe before turning fully south and then west in Pernand, with a few holdings facing north-west. In all there are 160 hectares of *grand cru* vineyard, yet only three appellations: Corton, Corton-Charlemagne and the seldom-seen Charlemagne. As a rule of thumb, when considering the wines, it is reasonably safe to assume that Corton is red and Corton-Charlemagne is white, save for a tiny amount of Corton *blanc*. Considering the vineyards is another matter, for the appellations overlap – if Pinot Noir is planted in the *lieu-dit* En Charlemagne, for example, the resultant wine is Corton.

With regard to planting, Pinot Noir finds favour on the mid- and lower slopes that face south-east and south, with Chardonnay prospering higher up, close to the tree line and in the vineyards that turn from south to west into Pernand-Vergelesses. There's a bewildering number of *lieux-dits* – over two dozen – and it would surely make sense to split the behemoth *grands crus* into these constituent parts and then rank them as appropriate. An obvious trio for designation as

30   CÔTE D'OR

grands crus in their own right would be Les Renardes, Les Bressandes and Le Clos du Roi, superbly sited as they are on the mid-slope with a south-east exposure. Above them lies Le Corton, another candidate for top rank, and not to be confused with plain Corton, which cannot be labelled as Le Corton: the definite article may only be used for wines that come from that *lieu-dit*. As it is, most of the wines are labelled with reference to their specific *lieux-dits*, Corton Clos du Roi, Corton Bressandes and so forth, unless they are blended across several, a reversal of the practice in Clos de Vougeot where the *lieux-dits* names are hardly ever used.

It is hard to generalize about the wines but too many reds hint at greatness by way of fleeting flavours without delivering: artists' sketches, not finished works. As with Clos de Vougeot, however, it is worth noting in general terms that the wines have gained more conviction in recent years. A good Corton-Charlemagne is a different matter. In youth it is tight-coiled and unyielding, steely not flashy, seldom lavish or lush, and only with age does it fill out and unfold to reveal a broad panoply of flavours.

South of the hill of Corton the communes of Savigny-lès-Beaune and Chorey-lès-Beaune face each other across the D974, Savigny to the west and Chorey on flat ground to the east, with a toehold on the 'right' side of the road. Savigny's *premiers crus* lie on slopes either side of the valley through which the little river Rhoin flows. Les Lavières and Aux Vergelesses are the best, well sited on the northern hillside, Lavières facing south, Vergelesses more easterly. Chorey possesses no *premiers crus*, a circumstance that works in favour of savvy consumers in search of good wines at reasonable prices.

The A6 motorway forms a rigid border between these two and Beaune itself, and it is hard not to feel that the Côte de Beaune proper only begins once you have crossed the main road. Beaune was dealt a generous *premiers crus* hand, as perusal of any vineyard map shows – their darker colour dominates, with attendant blobs of *village* above and below on the slope. The broad sweep is bounded by Beaune's suburbs to the east and hilltop forest to the west. At its heart lies Les Grèves, a roughly square block of 30-plus hectares that climbs the hillside from 225 metres at its base to 300 at its upper limit. To the north lie Les Toussaints and Les Bressandes and to the south Les Teurons and Aux Cras. Les Grèves, named for its stony soil, is the clear leader in potential – delivered on by the likes of Tollot-Beaut and Domaine Lafarge.

Côte de Beaune

In a region whose names are replete with complication and potential confusion this one tops them all. This appellation should not be confused with Côte de Beaune-Villages, nor does it refer to the southern section of the Côte d'Or, nor indeed does it rank as a principal vineyard. It refers to a scattered appellation of 66 hectares, spread across seven *lieux-dits*, located between 300 and 350 metres altitude in the hills above and to the west of Beaune. Some of the vineyards lie beside and to the right of the D970 road as you travel from Beaune to Bligny-sur-Ouche. The wines are seldom seen and it is included here simply because of its potential to confuse. 'Collines de Beaune' would surely be a better and certainly less confusing name.

A number of the large négociants have flagship Beaune wines upon which they lavish 'spoilt child' care and attention: Drouhin's Clos des Mouches, Bouchard Père et Fils' L'Enfant Jesus and Jadot's Clos des Ursules are examples. These wines stand on their own reputations while some of the lesser *premiers crus* are barely known outside the region.

The Pommard commune begins at the roundabout as you drive south out of Beaune on the D974 – fork right off the main road here to climb the gentle slope to the village itself, which sits plumb in the middle of its *premiers crus*. Then wriggle through the village to get to its most prestigious vineyard, Les Rugiens, divided by a small road into upper and lower sections – *haut* and *bas* – so much easier for English speakers than *dessus* and *dessous*. The name Rugiens derives from *rouge* and references the reddish soils caused by iron oxide, and it is the lower section that is most prized and frequently mentioned as a candidate for elevation to *grand cru*. Clos des Epeneaux is also mentioned whenever this long-running debate gains new legs but it is unlikely that anything will change in the near future. Of the other *premiers crus* Les Jarolières, adjacent to Rugiens-Bas, produces wine with a finesse not normally associated with Pommard, though recent warm vintages, allied to a more sensitive touch in the winery, have seen a lightening in the traditional Pommard style of heft before beauty.

As with Pommard, Volnay is solely a red-wine commune and is home to no *grands crus* either. There the similarities end. Pommard has considerably more appellation land, yet curiously the vineyards feel more expansive in Volnay, especially at the southern end, abutting Monthélie

32   CÔTE D'OR

and Meursault. Being further up the slope no doubt contributes to this illusion. It is at this end also that the acknowledged top rank *premiers* are found, Clos des Chênes, Les Caillerets and Taille Pieds. All three are evocatively named, none more so than the latter, a steep vineyard, the incline of which forced the vignerons to stoop low to prune the vines (*tailler*, to prune), so low that they risked cutting their feet.

Clos des Chênes and Taille Pieds are beautifully situated, straddling the 300-metre contour line on the map, yet it is Caillerets a little lower down that commands the greatest respect, where the stony soil yields the quintessence of Volnay. Other vineyards of note include Clos des Ducs and Les Santenots du Milieu, the chameleon vineyard that lies across the boundary in Meursault but which is labelled Volnay if the wine is red, Meursault if white.

The Côte d'Or stretches to its widest south of Volnay, spanning over 6 kilometres across from Saint-Romain to Meursault. The contrast between the two could hardly be greater, both in situation and renown. Saint-Romain is home to some of the côte's highest vineyards, some of which touch 400 metres in places and, if driving, first gear needs to be utilized as you pull up the hills and used again as you descend. As yet, none of the vineyards are ranked *premier cru* though moves are afoot to change this. There is a feeling of dislocation in Saint Romain, a sense of being well removed from the côte, though the reward comes by way of the scenic drive. The stony escarpment that forms the backdrop to the village adds a *sauvage* vista not found elsewhere. Saint-Romain may feel remote and rustic but in terms of beauty it can easily give the prim and patrician communes of the main drag, with their gentle manicured slopes, a run for their money. Next door, Auxey-Duresses' vineyards are strung along the course of the Ruisseau des Cloux, a small watercourse, with a handful of *premiers* at the eastern limit of the commune adjoining Monthélie. Here, the best vineyards face each other from opposite hillsides with the village between. Les Hautés, cheek-by-jowl with Meursault, produces a wine best described as 'junior Meursault'.

Corton-Charlemagne excepted, the Côte d'Or's finest white wine country starts at Meursault, the curiosity being that the commune is book-ended to north and south by anomalous vineyards that, if they produce red wine, are not Meursault but Santenots and Blagny. Only when planted with Chardonnay do they qualify for the Meursault appellation. Meursault is a big commune with a clutch of storied *premiers crus* such as Les Perrières, Les Genevrières and Les Charmes, though no *grands crus*. As if to compensate for that deficit it boasts a superior collection of *village lieux-dits* that in the hands of the best producers are regarded as *de facto premiers crus* and which are easily the equal of lacklustre *premiers* from elsewhere. A quartet – Les Narvaux, Les Tillets, Le Tesson and Les Vireuils – occupy favoured sites to the west of the village at about 300 metres altitude. The bulk of the *premiers crus* lie to the east and south of these, reaching as far as the boundary with Puligny-Montrachet. Named for old quarries, Les Perrières is the star, a fragmented vineyard of some 14 hectares that sits upslope of the smooth sweep of Charmes and Genevrières. In the right hands those two produce superlative wines, but Perrières can top them both by way of greater depth and insistence without losing its elegance.

Confused?

Vineyard names in the Côte d'Or can be wonderfully evocative – and sometimes mind-bendingly confusing. Clos de Vougeot Musigni is not a blend of two adjacent *grands crus* but the *lieu-dit* in Clos de Vougeot that is adjacent to Musigny. The final 'i' rather than 'y' is the clue. Côte Rôtie may be a famed wine of the northern Rhône but it is also a *premier cru climat* in Morey-Saint-Denis. The Pommard *premier cru* Clos Blanc only produces red wine. Petit Clos Rousseau in Santenay and Petits Epenots in Pommard are actually bigger than their neighbours, Grand Clos Rousseau and Grands Epenots, the explanation being that the rows of vines in the 'grand' vineyards are longer than in the 'petit' ones. And the Clos des Epeneaux *climat* covers part of the Grands Epenots and Petits Epenots *lieux-dits*. Remembering the difference in spelling between Les Encégnères in Chassagne-Montrachet and Les Enseignères in Puligny-Montrachet is never easy. Both derive from 'Les Sensennières' meaning the lands of the Sens bishop. And so on …

Elegance is a hallmark often cited for the wines of the next commune, Puligny-Montrachet, along with refinement, breed and poise. These are qualities seen to ultimate advantage in a great Montrachet, though its next-door neighbour a little higher on the slope, Chevalier-Montrachet, is not far behind when comparisons of outright quality are being made. Montrachet's situation is textbook perfect, a mid-slope slice of vineyard on a gentle incline facing south-east. Its 8 hectares divide almost 50:50 between Puligny and Chassagne, hence the appropriation of its name by both villages and, strictly speaking, the Chassagne section is Le Montrachet while the Puligny section does without the definite article, though this distinction is not rigidly applied. It hardly needs stating that the first 't' is silent, thanks to the conflation of two words, 'Mont' and 'Rachet', after which the pronunciation remained as if they were still two. Roughly speaking it means bare hill – it lacks the dense forest that runs along most of the Côte d'Or's hilltops. Montrachet is the most celebrated white-wine vineyard on earth and when on song the wine is indubitably magnificent but it can also leave expectations unfulfilled, especially when the price is considered.

In the same way that Chambertin has its satellites so too does Montrachet; in addition to Chevalier there's Bâtard-Montrachet, Bienvenues-Bâtard-Montrachet and Criots-Bâtard-Montrachet. Chevalier is slightly smaller than Montrachet and the slope is slightly steeper, with leaner soil that gives racier, less substantial wine. Bâtard's dozen hectares are shared almost equally between Puligny and Chassagne and it borders Montrachet on its western side, separated from it by a narrow road. Compared to Chevalier the soil is heavier, a distinction reflected in the wines: Bâtard plush and plump, Chevalier more clearly etched. Bienvenues lies wholly within Puligny and Criots wholly within Chassagne; separated by about

Looking across Clos de la Perrières to Château du Clos de Vougeot, with a sliver of Musigny in the foreground

500 metres, there's a yawning gulf between them in renown. Bienvenues comprises a block on the north-east corner of Bâtard with little to distinguish between them, and the wines are barely distinguishable also, while Criots slopes away from Bâtard on its southern side and is less favourably sited. Criots is the awkward child of the quintet and, for a *grand cru*, too often comes up short where it counts – on the palate.

Puligny is also home to some outstanding *premiers crus*, including Le Cailleret and Les Pucelles, both of which are contiguous with the block of *grands crus* and, in the right hands, capable of rivalling them for quality. A curiosity is the tiny amount of Puligny *rouge* that is produced, a distant echo from a time when Pinot Noir was widely planted in the commune. Indeed, with the exception of the *grands crus*, almost every *premier cru* and all but one of the officially recognized *lieux-dits* (over two dozen) are permitted to make red wine. If the vignerons chose to, they could convert nearly all of their vineyards to red and still call it Puligny.

Chassagne-Montrachet's vineyards form a reasonably cohesive rectangle with the *grands crus* at the top, abutting Puligny. These and a few others stand apart, split from the bulk of Chassagne by the D906 road that leads to Saint-Aubin. This road feels like the Chassagne–Puligny boundary but nobody in Chassagne is complaining that it is not, for there would be no *grands crus* in the commune if it were. As a consequence they are part of the commune but stand separate, like a choir balcony in a church. Save for the *grands crus*, every scrap of *village* and *premiers crus* vineyard may produce red or white wine. Some, such as Clos Saint-Jean and La Boudriotte, produce both side by side, which makes for interesting comparative tasting and discussion as to whether this or that vineyard is better suited to Chardonnay or Pinot Noir.

The stony Cailleret vineyard, beside the upper part of the village and facing south-east, is the leading *premier*, though Les Chaumées (not to be confused with Les Chaumes) on the boundary with Saint-Aubin can challenge it. And, in the right hands, Les Chenevottes can be excellent. To the south of the commune, Morgeot is a 54-hectare vineyard hold-all that contains well-known *lieux-dits* such as La Boudriotte and Clos Pitois and others such as Guerchère and Ez Crottes, whose names only ever appear on detailed vineyard maps.

Saint-Aubin sits west of Puligny and Chassagne, and lacks their compactness, projecting away from the Côte de Beaune in a dog-leg as the vineyards follow the varied slopes. Thanks to the jumble of those slopes the vineyards face in numerous directions, from northeast to south-west. Perhaps unsurprisingly, the two best *premiers crus* are found just around a turn in the hillside from the Montrachet *grands crus*. These are En Remilly and, above it at 350 metres, Les Murgers des Dents de Chien, the latter named for the *murgers* or heaps of stones piled at the edge of a vineyard, created by vignerons clearance work. Some are massive and prompt wonder at the toil that led to their creation.

South of Chassagne the Côte d'Or begins its long sweep westwards through Santenay to finish with a rustic flourish in Maranges. The first vineyard you meet – Clos de Tavannes – vies for top spot with its neighbour Les Gravières and perhaps Beaurepaire, which sits above the village, rising steeply from 250 to 350 metres. All three wines can age well, especially Tavannes. One of Santenay's biggest *premiers crus*, Clos Rousseau, sits on the commune's southern boundary and changes its spelling to 'Roussots' in Maranges. This, along with other Maranges *premiers crus* such as Le Croix Moines and La Fussière, are vineyards with little recognition outside their immediate locality but, as with the best of Marsannay right at the other end of the Côte d'Or, they are likely to become better known in the future.

More stone than soil in Les Gravières vineyard, Santenay

> ## Côte de Beaune-Villages
>
> Unlike Côte de Nuits-Villages this is not a site-specific appellation; no vineyards are designated as belonging to it. It's a catch-all appellation for reds only and the wines may come from fourteen different communes, stretching from Ladoix-Serrigny in the north to Maranges in the south. Those not included are Aloxe-Corton, Beaune, Pommard and Volnay. It's a handy appellation for quality-conscious domaines who may want to use it for the produce of young vines, while waiting for them to gain a few years' age before inclusion in the village appellation. Négociants can also use it when they blend parcels of wine from different communes. Wines labelled thus are simple and fruity, more Pinot Noir than Côte d'Or.

FROM ABOVE

Seen from the air, tracing its course southwards from Dijon, the Côte d'Or appears as little more than a ripple on the earth's surface, bracketed by dark green forest on the hill tops and, in late July, golden fields of just-cut corn on the plain. Flying at a couple of hundred metres above ground it is possible to appreciate the maze-like intricacy of vineyard boundaries, with the plots of vines interlocking like jigsaw pieces, laced with spindly roads and narrow tracks. You must maintain a fixed gaze, however; glance to right or left and there isn't a vineyard to be seen. Though expected, the tiny scale is the biggest revelation – the côte is diminutive yet brilliant. The colours are more varied than expected, vivid greens of different shades and intensity depending on the row orientation, interspersed with less verdant patches of fallow land. There are splotches of brown, too, where a vineyard is awaiting replanting.

Some villages are easier to identify than others and the relative sprawl of Gevrey-Chambertin signals the beginning of the great sweep of *grand cru* vineyards that crosses six communes. There is a blip as the village of Chambolle-Musigny interposes itself between Bonnes Mares and Musigny. Tucked into the rocky hillside, it might be missed were it not for its distinctive church tower.

The squat block of the Château du Clos de Vougeot sits within its *clos* and together they dwarf the adjacent eponymous village. The grand expanse of the *clos* is in marked contrast to the minute *monopoles* of

Vosne-Romanée vineyards from above, village bottom left. Romanée-Conti is the darker green block, centre, with 'notch' at top

Vosne-Romanée that follow. More than ever the tiny scale is apparent here – Vosne is not easily spotted and passes in a blink; the straight avenue of trees leading in from the D974 is a good marker. Nuits-Saint-Georges is a grand metropolis by comparison and looks like an urban wedge driven into the vine-clad slope. A thin wisp of vineyard then threads its way through and around stone quarries before the hill of Corton announces the start of the Côte de Beaune. As on the ground, the hill is the most significant feature on the côte, though its imposing ground-level form is reduced to an egg-shaped mound. From above, what's easiest to appreciate is the varied aspect of the vines, from east facing to west facing as they sweep around the hillside, receiving a different ration of sunlight depending on location.

A ribbon of blacktop cuts the côte north of Beaune as the A6, the *Autoroute du Soleil*, snakes through before turning hard right in search of southern sun. Beaune sprawls at the bottom of its slope of *premiers crus*, the incline looking almost flat from the air. Then Pommard segues into Volnay in a trice, the latter signalled by the broad front of Pousse d'Or and looking even more neatly wedded to a pocket in the slope than it does from the ground. Meursault is bigger than the surrounding villages and calls attention to itself by size and location, centred on a small hill and flagged by the tall *mairie* with its polychrome roof in an unusual combination of yellow, green and black.

Puligny-Montrachet and Chassagne-Montrachet come next, co-claimants to Montrachet. From the air everything written about it occupying the perfect position on the mid-slope resolves into clarity. Montrachet sits like the filling in a sandwich, an oblong strip, measuring some 600 metres by 130 and covering 8 hectares, with satellite vineyards on either side. If the vines represented seats in a stadium these would be in the corporate boxes. Chassagne is easily spotted, sitting slightly downslope from a quarry, and then the côte turns through Santenay and Maranges to a jumbled finish. Two markers highlight Santenay: the multicoloured roof of the château and the spire of the church, a tall slate needle. The topography is more varied here, less regular than further north, a mélange of topsy-turvy slopes, the hilltops touching 500 metres, with the odd vineyard up to 450 metres. It is easy to see the change in vineyard exposure as it swings from almost due east to fully south at Maranges, though there are individual exceptions on the irregular terrain.

Travelling by light plane, the Côte d'Or is traversed in a matter of minutes. From north to south it is embroidered with millions of vines and punctuated by villages where, tucked behind high walls, there are more swimming pools than expected.

GEOLOGY

In the simplest layman's terms, the Côte d'Or can be described as a tear in the earth's surface caused by vertical slippage that exposed a multiplicity of different types of limestone from the Jurassic period. In addition, the resultant gentle slope faces roughly east towards the rising sun, a combination that has proved ideal for the cultivation of the vine.

The whole area was once under a shallow inland sea and the climate was tropical. It was teeming with marine life, such as oysters and other shellfish, and as they died their remains settled on the seabed over the course of many millennia, gradually building layer after layer and eventually forming the limestone basis of today's côte. In time, the sea receded and about twenty to thirty million years ago the forces that created mountain ranges such as today's Alps also led to the formation of the slope we call the Côte d'Or. The côte forms the western edge of the Saône valley and generally lies at an altitude of between 200 and 400

40 CÔTE D'OR

metres. The D974 road roughly traces the divide between slope and valley; all the great vineyards lie west of it and there, the underlying bedrock sits just below the surface soil. In places some enthusiastic digging with a spade would soon reveal it whereas on the eastern side it could be over 100 metres beneath the surface.

The geological fault that created the côte acted like a cut across a multi-decker sandwich, whose layers have eroded and weathered to yield the complicated pattern of clay-limestone (*argilo-calcaire*) soil types seen today. The côte may have started life as a reasonably homogeneous slope but further complication comes from two other influences. The layers were compressed from north to south so that they have been pushed upwards in the Côte de Nuits and have sagged at the beginning of the Côte de Beaune, to emerge again at Santenay. Additionally, the effect of the last ice age about 20,000 years ago was to create breaks and irregularities in the côte – *les combes* – the side valleys that run roughly perpendicular to it. Out of these have spilt deposits of stone and soil that add further complexity to the côte's soils.

The most favoured vineyards lie in the midriff of the slope at an altitude of about 250 metres, where is found the best combination of drainage, soil, subsoil and exposure. The geology of each site imprints the same stamp on a wine year after year unlike, for instance, the weather or the winemaking, both of which can change markedly or merely by nuance. All other influences are mutable; changes in geology are measured in millions of years. Whether it imprints a stamp of minerality on the wine and whether that can be tasted, and used as a valid tasting term, is a currently raging debate in the wine world. The upper hand resides with the scientists who assert that the influence of minerals in the soil cannot be tasted in the wine. Yet minerality is a term I am happy to use and when others use it I know what they mean. It's a bedfellow of acidity but broader and less penetrating and less apt to produce a flow of saliva.

Whatever you feel about minerality, it is certainly possible to taste geological influence simply by sampling a range of wines from the same winemaker and same vintage, the only variable being the vineyard. It needs emphasizing that the Côte d'Or is a multi-faceted piece of land. The wines it produces could be likened to hundreds of pupils in a school; they all wear the same uniform but they are all markedly different. And by using single grape varieties that multiplicity of terroir messages comes through in a linear fashion, unobscured by the hand of the winemaker that is more evident in a blend.

It takes only passing knowledge of the côte's pattern, or lack of pattern, to understand why adjacent, superficially similar *climats* can yield radically different wines. The vineyard mosaic endows the côte with enduring fascination for wine lovers, frequently leavened with helpless frustration. It's a sweeping statement, but thanks to its geology no other wine region can engage and confound like the Côte d'Or.

VITICULTURE

The rigid symmetry of today's Côte d'Or vineyards, usually planted with 10,000 vines per hectare, a metre between rows and a metre between each vine (with some variations), is a relatively recent development in the history of the côte. Prior to phylloxera the vines were planted *en foule* – in a crowd – resulting in a much higher density and facilitating propagation by layering. This simply involved bending a shoot of a vine and burying it in the earth and once it had started to grow on its own account and develop roots it would be cut from the original plant. Phylloxera forced the grubbing up of entire vineyards and when they were replanted it was in the neat rows that still embroider the côte today. This facilitated the use of horses for ploughing, though their era was short, superseded by vine-straddling *enjambeur* tractors, then reintroduced today at some top domaines because they compact the soil less and, the jaundiced observer might add, are more photogenic.

Layering is no longer possible and so vineyards are now replanted using clones, plants derived from a single parent vine and propagated, or by massal selection where the vigneron takes cuttings from plants that are performing well and uses these for replanting. Clonal selection offers greater security by way of vines with a proven track record of reliability and resistance to disease; massal selection doesn't offer the same guarantee but maintains greater diversity in the vineyard. The phylloxera-resistant rootstocks onto which the cuttings are grafted have a significant influence on the resultant fruit. The favoured rootstock after the Second World War was SO4, which chimed with the zeitgeist of those times, when over-fertilization to yield big crops became the accepted practice. Recent generations of vignerons have rued their predecessors' favouring of this rootstock and are more mindful

42 CÔTE D'OR

than ever that replanting offers them the single greatest opportunity to influence wine style and quality for a generation to come. Most recently, with the extraordinary weather variability from year to year – perhaps drought one year with relentless rain the next – the question of which rootstock to use when replanting has become trickier than ever. When asked about this, more than one winemaker has agreed that what is needed is a 'Swiss-army-knife rootstock', something that can meet all the necessary criteria.

How long the land should be left fallow between grubbing up and replanting is a pertinent question, the decision being informed by the opposing considerations of benefit and cost. Two years is probably a happy medium, anything less not being effective enough and anything more being too costly. Once a vineyard is established, constant management is needed to get the most out of it. Weeds are now more likely to be controlled by ploughing or by growing grass between the rows, usually every second row, than by the application of herbicides. The grasses compete with the vines and thus control their vigour as well as helping to reduce erosion in steeper vineyards. Pruning is a tedious and time-consuming operation that is carried out in the early months of the year. Week after week the vignerons will be seen working slowly through their vineyards, often in bitter cold, with some warming sustenance coming from the homemade braziers-cum-wheelbarrows they use for burning the cuttings. The most commonly seen pruning system is *guyot simple* though *cordon de royat* is favoured for vines with a tendency to over-produce.

Organic (*biologique*) and biodynamic (*biodynamie*) viticulture are increasingly common in the Côte d'Or, particularly the former, whose basic principle is that grapes should be grown without recourse to the use of industrially produced synthetic fertilizers, insecticides and herbicides. The use of copper sulphate is allowed, though contentious, because, while the sulphur degrades quickly, the copper residue remains in the soil. Biodynamics takes things much further, both in terms of the philosophical beliefs behind it and the practical application of those tenets in the vineyard. Biodynamic viticulture derives from the teachings of the Austrian philosopher Rudolf Steiner, where the influence of the entire cosmos is taken into account when considering what treatments and practices are correct for the vine. Taken in isolation, some of those practices – such as the application of herbal infusions on days indicated by the lunar calendar – seem bizarre and the temptation is to dismiss the whole system as vinous voodoo. It is better to judge the results in the glass, however, and these are generally impressive. Whether that can be attributed to *biodynamie*'s efficacy or the talents of its practitioners is a difficult question to answer. It has some exceptionally capable winemakers as its proponents, all of whom share a common trait of rigorous attention to detail, a trait that would guarantee impressive results regardless of the philosophy they subscribed to.

> ## Design classic
> Take one old metal barrel and an equally ancient bicycle wheel. Slice open one side of the barrel lengthways and lay it into a frame made from metal tubing which at one end attaches to the wheel and at the other is shaped to form two handles, yielding a crude wheelbarrow or *brouette à sarments*. From late winter to early spring these are used as mobile braziers to burn the vine cuttings, as the winemakers prune their vineyards. Wisps of smoke trickle upwards seeming, on a calm overcast day, to tether the low cloud to the ground. It is a classic Côte d'Or vista, yet the *brouette*'s days may be numbered. They may fade away because of environmental concerns, as an increasing number of producers choose to mulch the cuttings for compost rather than send them skywards as smoke.
>
>
>
> *Design classic: a brouette à sarments seen in the Musigny vineyard*

44 CÔTE D'OR

Not all organic and biodynamic producers are certified as such, some citing the daunting burden of paperwork, others saying that they want to be free to apply an unapproved treatment if calamitous conditions, such as the threat of mildew in June 2016 or 2024, demand it. Both of these are valid points but there are undoubtedly vignerons who preach one or other gospel while practising as they please, happy to bask in the glow of right-on approval from consumers. Those same consumers should also remember that it is far easier to be an organic or biodynamic wine drinker than winemaker, to pronounce from on high that only those wines that meet the highest viticultural standards will pass one's lips. Such drinkers could do well to moderate their demands and not be too strident in their pronouncements. A chat with a winemaker forced to abandon organic practices, as some were in 2024, so as to save a crop and thus pay the bills, the mortgage and the wages would certainly help.

Ultimately, what can be achieved with even the most diligent vineyard management is limited by the ground's potential, and it is instructive to note the increasing complexity, depth and length when tasting through a conscientious producer's wines, from *village* through *premier* to *grand cru*. In the right hands the vineyard ranking makes perfect sense. Those hands will be busy throughout the year, combating every challenge, including the depredations of wild boar, deer and rabbits in those vineyards adjacent to the hilltop forests. Less easy to deal with are the tourists who trample about the celebrated *crus*, helping themselves to souvenir stones or bunches of grapes just before harvest, or the odd errant driver who careens off the road, destroying some vines before coming to an ignominious halt.

For much of this century a higher planting density, well exceeding the 10,000 standard, was the accepted norm amongst quality-conscious winemakers seeking to get the best from their vineyards. One apparent benefit is that vines planted thus deliver high quality fruit, more usually associated with older vines, at a younger age. In this pursuit, planting at 14,000 vines per hectare was not unusual and some went well beyond this, most notably Olivier Lamy in Saint-Aubin. More recently, however, some vignerons such as Véronique Drouhin, Céline Fontaine, Elsa Matrot and Pierre Vincent have suggested that a lower density may help to combat the challenge of climate change by reducing evapotranspiration through the plant material.

CLIMATE AND WEATHER

It was always said without qualification that Burgundy lay at the northern limit for the production of great red wine in Europe and, while that is still largely true, in decades to come the same assertion might be made for Alsace or Germany's Ahr or Pfalz regions. The forty-seventh parallel runs through the Côte d'Or, a fraction south of Beaune, placing it well to the north of Burgundy's traditional Gallic rival, Bordeaux. Being at the margin for proper ripening of grapes is an advantage and a disadvantage: the long growing season favours the development of subtle and complex flavours, while in a poor vintage the grapes might not ripen properly resulting in vapid, charmless wines.

The Côte d'Or has a moderate continental climate. Winters are generally cold and summers generally hot, but within that sweeping summary there can be great variation. Some winters never get properly cold, remaining grey and damp with little frost, and some summers muddle along with too much rain and not enough sun. Extreme conditions play their part too: spring frosts can wreak destruction on the nascent crop, while summer hailstorms can cause devastation, usually highly localized. Both of these scourges have struck with distressing frequency in recent years. In addition, extreme summer temperatures

After the hail. These are 'regular' size; they can be considerably larger

46 CÔTE D'OR

over an extended period can cause the vines to shut down, interrupting the ripening process, as happened in 2003, and heavy rain at harvest can cause the vines to suck up water, swelling the grapes and diluting the juice.

Whether climate change is responsible for the increased variability of the weather and the incidence of catastrophic frosts or storms is impossible to say, but leaving those aside most vignerons are agreed that its influence so far has been for the good, resulting in earlier and riper harvests. It is possible that the côte's climate is in a period of transition, with no certainty of what is to come. It may settle into a new norm or remain in the jumble we are now witnessing. If climate change leads to weather chaos then the recently surmounted challenge of reviving the over-fertilized, moribund vineyards may yet pale beside the challenges posed by it.

All of the above remains broadly true today but the 'jumble' might now be described as a patchwork quilt of different periods of weather, rather than a steady progression of seasons. Recent vintages, since the first edition of this book, have seen extreme heat, drought, incessant rain, spring frost and summer hail, warm winters and cool summers, very early harvests, and various 'once-in-a-lifetime' weather events that have come about with distressing regularity. In paraphrase, 'We could make a plan if we knew what to plan for,' has been the response from numerous winemakers when asked how they plan to cope with the ever-evolving climate of the Côte d'Or.

3

PRODUCING BURGUNDY

The winemakers of the Côte d'Or are commendably modest – though sometimes frustratingly so – about the contribution they make to the final product, keen to repeat the mantra of the moment that great wine is made in the vineyard. That rang true when first coined and, sincere as its expression still is, it has been recited to death since, ground into clichéd blandness. Initially it sounded profound; now it sounds trite. It would be truer and more enlightening to say that wine potential is set in the vineyard, that a line is drawn beyond which the final product cannot pass, no matter how dextrous the winemaker. By diligent work in the vineyard the winemaker sets the potential; the winemaking challenge is to realize it. However, trying to exceed the potential by way of over-extraction or using too much new oak, for example, is simply using brute force, bolting on extra flavour components that the wine cannot carry. Pushing to exceed the potential results in misshapen wine, impressive at first but charmless eventually.

Which brings us to today's mantra: 'infusion not extraction'. In short, this recognizes the sins of the past while also acknowledging that with warmer vintages and riper grapes there is no need to push for extra colour and flavour. It is there waiting to be tapped and must be tapped in a sensitive fashion. All it now takes to get sufficient from the grapes is some clever coaxing; there is no need to metaphorically whack them to give up their goodness. Countless winemakers the length of the Côte d'Or now sing the infusion-not-extraction song, meaning that it is well on the way to becoming as hackneyed as 'great wine is made in the vineyard'.

48 CÔTE D'OR

Producing the best fruit possible in the vineyard cuts down the need for manipulations in the winery. It's like a tailor cutting cloth for a suit: get the cut right and few adjustments will be needed later. There will usually be enough vintage-related issues to be dealt with and it is the winemakers' ability to meet these challenges that separates the great from the ordinary. Those blessed with the magic touch are distinguished from the journeymen by their instinctive feel for what is necessary and their ability to wrest impressive flavours in unpropitious circumstances. Real skill is demonstrated in not reaching for too much, as in the trio of warm vintages 2018, 2019 and 2020. The potential was there, oodles of it, but a sensitive hand was needed to harness it. Just as a great chef is often marked by what is left out of a dish, so too the great winemaker is the one who knows when not to interfere, which in itself is a crucial decision. The role is both active and passive, and the best vignerons don't use a winemaking-by-numbers approach; they are alert to the wine's needs. In a vintage with abundant potential those dazzled by the siren call of making the wine to end all wines can trip themselves up with over-opulence at the expense of harmony and balance. Standing back is always the hardest decision because it involves inaction and passive observation rather than action and engagement.

THE GRAPES

Pinot Noir and Chardonnay account for almost all the wine produced in the Côte d'Or, and another pair of lesser though increasing renown, Gamay and Aligoté, account for most of the rest. The famed pair are the varietal pillars upon which the reputation of Burgundy stands. Nobody would argue, however, that if only one grape was picked as the standard bearer for the Côte d'Or it would be Pinot Noir. Where Chardonnay is exalted, Pinot is revered to the point of worship, perhaps like no other grape on the planet. In the wine world Chardonnay is ubiquitous, producing serviceable wines almost everywhere it is grown and superb ones, good enough to rival the Côte d'Or's best, in some selected regions. The same is not true of Pinot, poor expressions of which range from weak and vapid to overcooked and jammy. There are some excellent examples from the likes of New Zealand, Australia, South Africa, the United States and Germany, but Pinot Noir from a favoured site on the Côte d'Or, a good vintage and a talented winemaker remains unsurpassed. 'But for how long?', some might ask. A combination of warmer vintages, such as 2018, leading to heartier burgundy, allied to continuing improvements in the wine nations named above, may yet tilt the balance in favour of the newcomers, though probably not for a few decades yet.

Pinot Noir

It is almost impossible to avoid the word 'fickle' when describing Pinot Noir and the challenges it presents to winemakers. 'Capricious' is a good substitute, as are 'infuriating', 'mercurial' and 'temperamental'. Indeed, a whole thesaurus could be thrown at it without capturing Pinot's two-sided nature; it enchants and frustrates in equal measure, calling to mind a tempestuous love affair. Notwithstanding these difficulties it possesses a near-mythical appeal, a rite of passage attraction that suggests you haven't arrived until you have made Pinot. Perhaps 'diva-esque' captures its nature best: the soprano who can reduce an audience to tears with an aria – or a last minute cancellation.

Pinot Noir is thin-skinned both literally and figuratively, a quality that carries with it myriad potential problems. It doesn't like too much sunshine, or hot climates where it will ripen too quickly resulting in wines with baked flavours like overcooked jam. A cooler climate and slower ripening is essential for getting the most out of Pinot. The light skins also make it more susceptible to diseases and more prone to rot. Those skins also carry less colour and tannin – not in themselves problems – but in a world gone mad for depth of colour it might tempt the vigneron down the path of over-extraction, a course that can knock Pinot's signature delicacy on the head. A violet colour in youth that resolves into vivid, almost lucent, crimson with some age seems right. Being lower in tannin it relies more on acidity for backbone and age-ability.

Pinot's character is usually described in terms of fruits such as cherries, raspberries and strawberries but this is a simplistic template, the background wash upon which the vineyard, vintage and vigneron cast their colours. In the Côte d'Or Pinot acts a transmitter, carrying the minutely differentiated messages from vineyards that may be adjacent and superficially similar. As such it is probably the most sensitive of all grape varieties, echoing subtleties that others would mask. Notwithstanding the delicacy for which Pinot is renowned the wine is not a lightweight: there's depth, intensity and length. It should not be massive; if so, the flavour becomes lopsided with too much superstructure and not enough foundation. Yet the range of styles it produces in the côte is extraordinarily diverse. Pinot dances to many tunes there and any description of the grape's qualities when referencing its use on the côte risks being too rigid. Perhaps Chambolle-Musigny and Volnay most closely resemble the widely accepted stereotype – which doesn't capture a great Chambertin or a four-square Pommard. The range of styles guarantees enduring attraction.

A varied landscape matched with a grape of diverse character has for centuries yielded wines that reach an unrivalled peak of magnificence, wines of scent and substance and haunting beauty. Almost any lengths will be gone to in an effort to understand them. Pinot unlocks the terroir's potential, releases the genie from the ground. It can come out mute or singing like a nightingale and it is the winemaker who dictates that by dint of clumsy or dextrous winemaking. There's an indefinable whiff, a slightly untamed character to Côte d'Or Pinot, something that has remained wild despite the intensive viticulture and meticulous attention to detail in the vineyards, followed by equally fastidious practices in the cellar. Despite all this the *sauvage* element remains and that, for me, is what gives red burgundy its unique character and enduring attraction. It will always be unique.

Hidden burgundy

Red Chassagne-Montrachet is hidden burgundy, known to locals and well-informed consumers but largely overlooked otherwise. It doesn't have the outright class of Côte de Nuits reds, the weight and structure of Gevrey-Chambertin or the satin texture of Chambolle-Musigny, yet it possesses an immediate likeability, courtesy of vibrant fruit flavours, that makes it hard to ignore. It can also age much better than might be expected for a wine of immediate appeal without overt power. As an introduction to red burgundy there's hardly a wine to match Chassagne rouge. Time was when it was a bit coarse, a country-cousin red, but today sees greater finesse and precision of flavour; the rustic rasp is still there but it doesn't dominate the way it once did. Red Chassagne remained hidden because the 'Montrachet' in the name suggested this was exclusively white wine country. It deserves greater recognition, and is beginning to get some, but it remains the preserve of savvy consumers.

Chardonnay

Was ever a vinous name so abused as that of Chardonnay? It started life simply as the name of a grape (and a village in the Mâconnais, south of the Côte d'Or) but in the final years of the last century the word was dislocated from the grape and came to represent a style of wine: heavy and rich, high in alcohol, replete with tropical fruit flavours, over-oaked. The wines were lumbering and ungainly, like steroid-boosted gym freaks. Those were the 'high-end' wines; at the other end of the spectrum came unremittingly bland libations whose Chardonnay character was no more discernible than a watermark on a bank note.

'Anything but Chardonnay' became the catch-call of a new generation of wine drinkers who found refuge, if not much delight, in bland Pinot Grigio or perky Sauvignon Blanc. And if wine neophytes expressed a fondness for Chablis and a dislike of Chardonnay they were wiser than credited. Granted, they spoke in strictly factual error but their perceptions were accurate. White burgundy, made from the Chardonnay grape, could never be mistaken for 'Chardonnay', the wine style. At the same time Burgundy wasn't doing much for the Chardonnay cause either, thanks to the alarming incidence of faulty white burgundies caused by premature oxidation (see Chapter 7).

Despite these challenges Chardonnay has survived thanks to a set of attributes that see it producing the greatest dry white wines on earth at one end of the quality scale and serviceable ones at the other. In short, Chardonnay is easy to grow and its wine is easy to make. It is vigorous and prolific though not problem free. It needs to be held in check by way of severe pruning or else the large crop will be low in acidity, as the grapes will be if they are allowed to over-ripen, when the flavours turn to melon, pineapple and mango and the texture turns luscious and even oily. It buds early leaving it exposed to the threat of spring frost and never was this seen more tellingly on the Côte d'Or than in late April 2016 when the temperature dropped overnight and froze the buds, which were then burnt by the rising sun the following morning.

A critical decision for Burgundy's winemakers today is when to harvest, and the current trend is to pick earlier in the quest for lean not lavish wines. Within that trend lies an ever more important decision, as harvests across the board start earlier and earlier: the precise picking date. A sunny day in late August has a greater ripening effect than a sunny one a month later; the grapes can race towards optimum ripeness (as defined by the winemaker's philosophy), meaning that a day one side or the other of ideal can carry a heavy penalty in terms of how closely the finished style of the wine matches the desired style. But should white burgundy be lean? Or rich? Can it be both? It can certainly be strong without being hefty. For example, a great Chevalier-Montrachet is subtle and nuanced with deep layers of flavour and resonating length on the finish. It is vigorous rather than overtly powerful and, in addition to ripe fruit, there is a savoury element, a faintly saline cut that adds complexity and freshness, making repeated sipping a lively, revealing exercise and not a trial for the palate. As with all trends, the move towards greater leanness will eventually be taken too far and the wines will end up shrill. Then the pendulum will begin its slow swing back.

Minerality

Is there a more divisive tasting term than 'minerality'? Some use it freely, without recourse to elaborate qualification, while pedants like to point out that vines cannot draw minerals from the earth, hence the term is invalid. But many tasting terms, widely used and understood, crumble to nothing when subjected to an analytical gaze. No one raises an eyebrow when we call a wine dry, though it is obviously wet, or white when it is clearly yellow. We know what we each mean. It is simply indicative of the struggle to capture a flavour in words, rather than an attempt to use language disingenuously or give erroneous information. The whole tasting vocabulary collapses if an overly rigorous interpretation of its meanings is applied. It is an accepted argot.

Aligoté

We have entered the Aligoté age. Continuing to cast it as Burgundy's awkward child, whose razor-wire acidity could only be ameliorated by a splash of crème de cassis, as per the recipe of the famed Canon Kir of Dijon, is now hopelessly outdated. Following the canon's dictum was like giving sweets to a cantankerous child, it was little more than a short-term mask for the underlying problem. Not any more, a haughty response to the offer of a glass of Aligoté belongs to the past.

It is still possible to find examples that cleave to the old template but increasingly Aligoté is being made by many of the côte's most talented vignerons in a fresh and lively style that would be ruined by any additions. They pride themselves on the quality of their most basic wine and it is almost a badge of honour to get it right.

It is usually relegated to less-favoured sites though there was a time when Corton-Charlemagne was planted with it. When properly ripe the signature acidity is ameliorated by the fruit and although it can age well it is usually drunk young. It can only be a matter of time before an ambitious winemaker takes the bold decision to plant Aligoté in place of Chardonnay on a prime site. Indeed, such an example already exists in the shape of Domaine Ponsot's Morey-Saint-Denis *premier cru* Monts Luisants, which ages magnificently. Who knows, by century's end it may be the Côte d'Or's principal white grape.

Gamay

Gamay is the grape of Beaujolais and suffered the indignity of being banned from the Côte d'Or by Duke Philip the Bold in 1395. Judging by the intemperate language used in the wording of the ban he may have had a bad experience with Gamay the evening before. It was branded as harmful to humans and being possessed of dreadful bitterness and anybody who had it planted was given five months to rid their vineyards of it. Subsequent centuries saw the ban repeated and it is not fanciful to argue that its reputation has never recovered.

It may have been named after the village of Gamay, which is passed when travelling from Chassagne-Montrachet to Saint-Aubin, and there is still some planted thereabouts. Apart from that there is precious little to be found on the côte and that tiny amount is usually blended with Pinot Noir to make Bourgogne Passetoutgrains. It is not a profound wine yet it is significant that it is made by Michel Lafarge and was also made by the late Henri Jayer, two of the côte's most esteemed vignerons. As with Aligoté, Gamay is capable of ageing well but because of its freshness and the immediate singular appeal it is seldom given the opportunity. An interesting recent development is the number of highly regarded Côte d'Or vignerons who are now making wine in Beaujolais, some of exceptional quality. If anything can restore Gamay's standing and right a 600-year wrong, this can.

HARVEST

The Côte d'Or is a magical place at any time of the year but there is no denying it is even more so at harvest time. Through the summer

54 CÔTE D'OR

there will be mounting speculation as to the quality of the harvest and, as August progresses, debate as to its likely start date. There used to be a starting gun in the form of the *Ban des Vendanges*, an official declaration before which nobody was allowed to pick. It was a throwback to pre-Revolutionary times when the local lord was permitted a head start on all others, who had to wait for the *Ban*. It continued in use until the early years of this century, as a means of preventing growers in a hurry from harvesting unripe grapes, but the heatwave summer of 2003 caught everybody napping and it has since been abandoned.

Deciding when to pick is probably the most important decision the vignerons will make every year and over the course of a career they may get no more than forty chances to call it correctly. The decisions, good and bad, will crucially influence the quality of the wines and will be locked in them until the last bottle is drunk. Getting it right is paramount. Optimum ripeness is what vignerons are after, which sounds simple enough until a consideration of what is optimal leads in more than one direction. A simplistic definition of ripeness is when enough sugar has developed to convert into a desirable level of alcohol, with enough acidity to retain freshness in the finished wine. Yet if that is the only measure applied then no account is taken of tannins and colour components. If tannins are not ripe the resultant wine will be raw and rasping, instead of pleasantly crisp. All of this applies in other regions but in the Côte d'Or it has added significance because wines made from under-ripe or over-ripe grapes tend to taste only of the grape variety, with little transmission of the vineyard character. The terroir message is like a signal that only comes through on a narrow frequency band, so getting the harvest date right is crucial. If it is missed the wines lose their fabled Côte d'Or calling card – the fact that they carry with them a sense of place and not just the flavour of the grape.

There will be more prosaic considerations pressing in on the decision making: in crude terms the harvest represents the annual income, from which wages must be paid, repayments met on bank loans and living expenses covered; perhaps the grapes are not ideally ripe but rain is forecast so maybe best to harvest now; or there's a team of fifty pickers going to remain idle if picking is suspended for a day in mid-harvest.

Most Côte d'Or domaines own a spread of vineyard parcels so a decision as to what order to harvest them in may be dictated by different ripening rates, which helps in getting them all harvested as close to optimum as possible. How the operation itself is performed can vary from domaine to domaine. The moral high ground of harvesting is held by the experienced team of pickers who return to the same domaine year after year and are fed and housed there, welcomed as old friends in fact. Many of them take their annual holidays to do so and for a conscientious vigneron they are an indispensable aid to achieving top quality. Other domaines have switched to an easier to manage model employing an agency to provide a specified number of pickers on a specified date, together with packed lunch. Making sure they do a diligent job is not as easy as instructing a twenty-year veteran who probably knows the vineyards as well as the owner. The odd mechanical harvester is seen too, but whether their touch is as dextrous as an experienced picker's is questionable. A good picker is the first step in the selection process, the first filter the grapes must get through, someone able to reject rotten, damaged or unripe bunches as necessary.

Without getting dewy eyed, it is instructive to reflect that the essence of harvest – getting grapes from the vineyard to the winery – has not changed for centuries. Docile villages bustle into life as hordes of pickers, *vendangeurs,* descend on the Côte d'Or and a collective frenzy

Harvest 2020, 24 August at 14.43. These grapes were picked within the hour

grips the region, the measured pace of life abandoned. Early morning sees them assembling in the villages awaiting transport to the vineyards while devouring croissants and cigarettes in equal measure. The hillsides come alive as they move through the vineyards, the green sward pricked with the multi-coloured dots of their clothing, and a host of little white vans dart hither and thither, driven by anxious vignerons supervising the harvest from their scattered vineyard plots. One day they are joined by another group, cycle tourists equipped with helmet-mounted cameras and clad in vivid Lycra, who zoom about, filming and photographing. It all ends with a rambunctious party at each domaine, before calm descends again for another year.

WINEMAKING

'It depends on the vintage' is a frequent response when winemakers are asked about their methods and philosophy. A cynic might dismiss this as canny evasion of probing questions but the abandonment of the one-size-fits-all approach of yesteryear is a more valid explanation. The philosophy is no longer written in stone. Today's vignerons are more responsive to the needs of each vintage and less likely to impose a rigid policy inherited from a previous generation. Dogma plays little part

Post harvest, November 2024 – the grapes that got away

today and specific facts and figures about vinification that carry spurious certainty are not as readily trotted out as previously.

The flexibility of approach is not solely in response to vintage conditions but also respects vineyard variation, how the grapes from one plot are different from another and so require different handling in the cellar. Origin, which begets identity, is cherished like never before. If Côte d'Or burgundy does not taste of where it comes from then it loses a major part of its identity, the sense that this wine can only be made in that place. In the closing decades of the last century an international style of wine, often based on Cabernet Sauvignon, gained great currency worldwide. The wines were impressive, well made, beautifully polished with compelling flavours, but whether they tasted of where they came from is debatable. They could have been made anywhere. Their character was formed in the winery rather than the vineyard. They were usually enjoyable but seldom exciting.

Who made you?

This is the first question to ask of a Côte d'Or wine when considering a purchase. Nothing matters more and until the answer is known almost all other information is irrelevant. Vintage and vineyard can be ignored until it is ascertained whose hand crafted the wine. Is that hand skilled and committed or clumsy and lazy? Does it belong to an ambitious young name on the rise or to one that is happy to coast on a reputation established by a previous generation? If the label says Gevrey-Chambertin or Volnay we can glean some idea of the wine's style but the best indicator of quality is the producer's name. Where terroir is nature, winemaking is nurture. When there is synergy between the two, great wine results. Sleuth-like precision must be applied to getting a name correct. There are numerous iterations of 'Colin' and 'Morey' in Chassagne-Montrachet and Saint-Aubin. Just as the property buyer's mantra must be 'location, location, location' so too the burgundy buyer's must be, 'producer, producer, producer'.

A sense of place is the Côte d'Or's trump card, overplayed at times, but valid nonetheless. Winemaking that does not respect that, winemaking that is too influenced by the globalization of wine styles and consumer tastes, risks obscuring it and ultimately losing it. The pressure to conform to a standard template is a hard to resist siren call, but

58   CÔTE D'OR

if Côte d'Or winemakers answer it they will be on a one-way street to decline. If the unique quality and variety with which the côte imbues Pinot Noir is lost through homogenization of style then it will taste like a wine from anywhere – and everywhere can make it more cheaply than the Côte d'Or.

At present, winemaking there is an individual affair practised by hundreds of artisans; instinct and knowledge are co-workers, backed by experience. The best Côte d'Or wines are the tailor-made suits of the wine world: no two are the same, each carries the stamp of whoever 'stitched' the final product. Quirks and foibles add character, too much at times, but the polished perfection of mass production, admirable rather than lovable, is absent here. And that is the way it must remain.

RED WINE

Arriving at the winery, it is likely that the grapes will have undergone some degree of sorting at the vineyard, particularly to remove rotten bunches or parts thereof. It is now established practice that they are then sorted on a sorting table where a team stands either side of a slowly moving conveyor belt removing sub-standard fruit, usually rotten but also perhaps under-ripe or damaged, or even grapes that have been shrivelled dry by the sun. Vibrating tables are also used where the bunches are riddled and any sub-standard berries fall off. Optical sorters are also used in the Côte d'Or, programmed to identify faulty fruit and to remove it with an accurately directed jet of air.

The sorting tables come into their own in difficult years when, for instance, hail damage may lead to rot – a whole cuvée could be tainted by the inclusion of a small quantity of rotten grapes. It's like adding a single mouldy grape to a fruit salad, it pervades and degrades the whole bowl. Rotten grapes do the same to wine; their flavour is like an ink stain on a white shirt. Sorting to remove rotten and damaged fruit is one thing but sorting to include only the best, most perfect grapes is another. Some producers apply extremely rigorous standards to their sorting; perhaps it is on the way to becoming too rigorous, too intense in the search for perfection. If it does, they risk producing 'beauty queen' wines, polished and perfect, but lacking attraction and character, making them difficult to engage with. It has yet to happen, the wines are still characterful, and I hope it never does.

Stems

No single winemaking topic prompts as much debate in the Côte d'Or as the subject of stems and whether to use them or not. Every winemaker has a personal preference, which will be cogently, sometimes trenchantly, enunciated. Traditionally they were used for the simple reason that removing them was not the easy task it is today. Traditionally too, it meant that much red burgundy had a coarse streak of flavour running through it thanks to the inclusion of unripe stems in the fermentation. De-stemming increased in popularity in the decades following the Second World War mainly thanks to its being practised by Henri Jayer, one of the most celebrated winemakers Burgundy has ever seen. Jayer was the high priest of 100 per cent de-stemming, and he made wines of such renown that his influence was immense. His wines enjoyed an extraordinary following, thanks to their wonderful intensity of fruit and graceful length on the finish, and they continue to fetch prices at auction that rival Romanée-Conti. With a winemaker of his standing championing the practice it was inevitable that it caught on.

The point was reached where it became almost *de rigueur* to de-stem completely, though some prestigious domaines remained notable exceptions, such as Romanée-Conti and Dujac. All trends wax and wane – the use of new oak barrels for ageing is the best example – and what looked like a fixed position a few years ago is more fluid now. Today, along with a reversion to more traditional practices in vineyard and cellar – ploughing with horses is the most camera-friendly – there is a trend towards using more stems, the previous polar positions of all stems or none have softened and, in richer vintages such as the warm trio of 2018, 2019 and 2020 many vignerons will use a percentage of stems so as to lighten a wine that may otherwise end up too rich.

It is agreed that on the plus side stems absorb alcohol and draw heat out of the fermenting must, and their addition also results in a paler wine. The downside is that if the stems are not properly ripe they can add an unwelcome flavour variously described as 'bitter', 'green', 'astringent' and 'vegetal'. In the finished wine, whole bunch fermentation produces tactile results, it adds a covert quality, some background ballast, a textural smoothness that fills and rounds the wine. De-stemming, on the other hand, endows the wine with a bountiful fruit character, racy and pure, yielding a more singular impact on the palate than the broader imprint of the whole bunch version.

Fermentation and élevage

Once the grapes are in the vat a cool soak of about four to five days usually follows. Tannin and colour are thus extracted, but the biggest benefit of holding the must cool before fermentation begins is the extraction of attractive aromatics. Once fermentation is underway the solid matter rises to the top and forms a cap that would remain separate from the must below, thus retaining valuable colour, tannin and flavour, if it weren't regularly forced back into contact with it by carrying out *pigeage* and *remontage*. *Pigeage* (punching down) involves plunging the cap down into the must by hand with a paddle or, in larger domaines, with modern hydraulic apparatus. *Remontage* (pumping over) sees the must being drawn from the bottom of the vat and then poured or sprayed onto the cap. This also aerates the must.

Most vignerons will use a combination of both to effect the desired extraction. Depending on vintage and vineyard they may favour one process over the other, and may also use one or the other at different stages in the fermentation. *Pigeage* is the age-old process and used to involve the vignerons stripping off and jumping into the vat to use their body weight to do the job. It is the rougher of the two processes and may result in harsh flavours – from crushing the pips, for instance – while *remontage* is less aggressive and the ease with which it can now be carried out with modern equipment has seen it become more widespread in recent decades.

A procedure that has become much less common is that of chaptalization, adding sugar to the fermenting must to increase the alcohol level in the finished wine. It was once widespread, undertaken every year with hardly a thought for whether it was necessary or not. It usually was, because of the abundance of unripe vintages, and in those circumstances it didn't simply boost the alcohol level, it also acted as vinous make-up, giving the wine a better texture and papering over the cracks, as it were. Such a heavy-handed approach was understandable but today better quality fruit, thanks to better vineyard management, and more consistent ripeness thanks to warmer vintages have largely obviated the need for chaptalization. When used now it is done with greater discretion and may involve the addition of several small amounts of sugar rather than one large dose, the primary aim being to extend fermentation and thus gain better extraction, without boosting the alcohol level by much. Tartaric acid may also be added to the wine if it is considered deficient in that respect, but added acid doesn't always integrate fully into the wine and can strike an errant note on the finish, jarring an otherwise pleasant flavour.

Other vessels

In the last decade there has been a noticeable increase in the use of non-traditional vessels in the winemaking process. The rigid regime of vinification in stainless steel or wooden vat followed by élevage in 228-litre oak barrels is no longer *de rigueur*. The first sign of change was the use of larger barrels to reduce the oak impact. Then followed clay amphorae and concrete eggs for fermentation and, more recently, glass globes for élevage. Stainless steel has also been shaped into large eggs and even barrels. Ceramic containers of varying shapes and sizes, while not widespread, are no longer an unusual sight. It is too early to say what effect all these changes will have but one thing is certain: oak influence is on the wane.

After the wine is run off into barrels, or other vessels, the remaining solid mass, the *marc*, is pressed to extract the concentrated press wine, which is then usually added to the free run. Malolactic fermentation ('malo') follows, softening the acidity, rounding out the hard edges so that it sits better in the overall flavour of the wine, rather than standing separate. Racking usually occurs after malo to remove the wine from its lees and there will probably be another racking before the wine is assembled to rest in tank for a few months before bottling. Whether it is fined and/or filtered at that stage will be a decision for the winemaker. A consumer craze for wines that have undergone neither manipulation has seen both cast as villainous intrusions on a wine's sanctity. That's an extreme view; done with a sensitive hand both processes are unlikely to harm a wine. On the other hand, aggressive filtering is like shaving someone's head – it changes their appearance radically. And, in the case of the wine, the 'hair' doesn't grow back.

WHITE WINE

It is almost impossible to consider white winemaking in the Côte d'Or without mentioning the problem of premature oxidation, which is dealt with in Chapter 7. Leaving that aside, white winemaking is generally a

62 CÔTE D'OR

simpler operation than red, starting with the sorting process, which is usually less rigorous than the near berry-by-berry approach now practised by some vignerons for their red wines. After that, whole bunches normally go straight to the press, though some vignerons give them a light crush first. Nowadays the presses are nearly all pneumatic though some domaines continue to use the traditional vertical press, which is operated hydraulically, and some others are changing back to them. Then the juice is allowed to settle in tank for perhaps twenty-four to forty-eight hours and may also be fermented in tank, though fermenting in barrel is generally considered to result in a better marriage between wine and wood.

During élevage the most marked difference between white and red winemaking is the use of bâtonnage or lees stirring, carried out by inserting a long, thin stainless steel implement, like a large bread dough hook, through the bunghole in the barrel and agitating the wine with it. The wine feeds on the lees, gaining texture and richness. Each domaine will have its own bâtonnage regime, perhaps once a week or once a fortnight, but the trend today is for less, not more. Malolactic fermentation is standard practice though in a ripe vintage some vignerons may decide to stop it early or indeed prevent it from occurring at all, so as to retain freshness in the wine.

However important acidity is to red wine it is even more so to white. Red gains its structure from tannin, while also relying on acidity to add balance and longevity. White relies solely on acidity to give it lift and length. It is not so much that it is more essential in white wine but that it has a greater role to play and because of that there is a greater risk of it being exposed by clumsy winemaking. Acidity is a white wine's central nervous system, its vital force, but it must be properly integrated so as to do its job unobtrusively. Thus adding acidity to white wine runs a greater risk of warping the flavour profile than with red. The initial impression of such a wine may be perfectly pleasant, with a nicely streamlined flavour – and then the acidity pops up on the finish, attached to the wine rather than part of it, like stabilizers on a child's bicycle, holding the wine upright when it might otherwise wilt.

Come bottling time, the vigneron is faced with a decision that is now considered critically important for white wines: what closure to use. Problems with TCA taint (2,4,6 trichloroanisole) affect all wines but premature oxidation is a whites-only malady. Corks, how they are treated and the lubricants used on them, have been the subject of intense scrutiny to see if they were the cause of the problem. As a consequence vignerons are exercising much greater diligence in sourcing their corks and insisting on much higher specifications. Some have switched to screwcaps while others, including a number of high-profile domaines, have abandoned corks in favour of the composite DIAM closure.

IN CONCLUSION

Côte d'Or winemaking in the broadest sense, from planting to bottling, is like making music. The ground is the musical score, the vine is the instrument and the winemaker is the performer. The hundreds of *climats* can be likened to hundreds of scores and the better the selection of rootstock and vine the better they will transmit the score. With instruments, a Stradivarius or a Steinway will transmit the music better than a more modest instrument. How it is interpreted is the performer's task. Just as the performer must be responsive to the score, so too the winemaker to the *climat*. And, just as the pursuit of flawlessness in performance is mistaken, favouring technique over interpretation, so too with winemaking. Flawless wines tend to be appreciated rather than loved, they are sterile evocations of something that should be vibrant and engaging. The soil is composed of a dazzling array of components, just like the score. But it sits mute, its message needs a conduit and a facilitator: vine and vigneron.

Different winemakers, all highly skilled, will produce markedly different wines from the same *climat*, just as performers will interpret the same piece of music differently, though the threat today is that they all play the same way, imitating stars in their field, and the threat in winemaking is that an increasingly international template is being imposed on wines no matter where they come from. Wines that result from this approach dazzle on first acquaintance but they are all shout and no echo, there's no depth, their fascination doesn't endure. Sensitive winemakers will respond to the unique set of circumstances pertaining to their grapes, rather than slavishly following external prescripts, to deliver subtle wines that rely on grace, elegance and intensity to make an impression. In the case of the Côte d'Or subtle should not be confused with simple; the côte is the birthplace of some of the most complex wines on the planet.

4

THE VILLAGES AND PRODUCERS OF THE CÔTE DE NUITS

Travelling south out of Dijon, the grandeur of the historic centre gives way to a suburban sprawl of car showrooms, supermarkets, hotels, fast food restaurants and the like, before vineyards begin to assert themselves around Marsannay-la-Côte, where the Côte de Nuits begins. It forms the northern section of the Côte d'Or and is home to the greatest red wines of Burgundy, though such a prosaic statement fails to convey the renown in which this strip of vineyard and its wines are held in every corner of the wine-drinking world. No collection of superlatives adequately captures the impact a great Côte de Nuits red makes on the palate and etches on the memory. Mere words are pedestrian; words set to music in a great operatic aria come close to matching the all-enveloping, sensual impact of these wines.

Walking the vineyards, passing Chambertin, or standing at the miniature T-junction where Musigny abuts Clos de Vougeot, or looking back towards Vosne-Romanée from the gaunt cross that marks the Romanée-Conti vineyard, it is not possible to imagine the stir these names cause in certain circles. Even if prices at the very top have softened a little in recent years, demand for the most illustrious wines seems insatiable. Witness the frenzied excitement that greets the arrival in the auction room of a prized parcel of old wines from a celebrated producer.

There is a patrician feel to the Côte de Nuits that is largely absent in the Côte de Beaune. The top *grands crus* engender a reverence that,

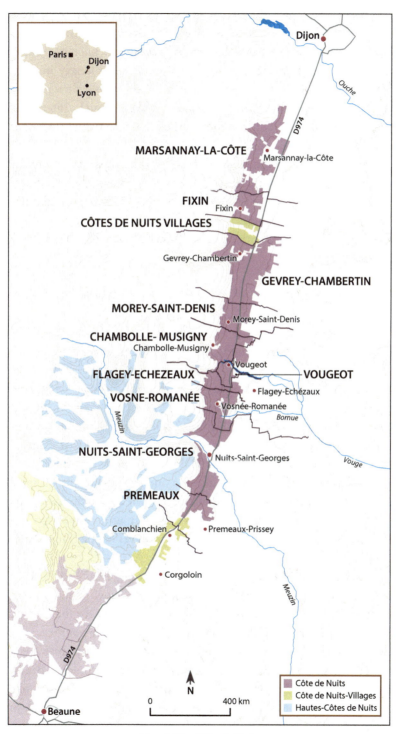

Map 1: The Côte de Nuits

apart from Montrachet, the Côte de Beaune does not match. That these fabled vineyards sometimes do not live up to the hype doesn't seem to dent people's enthusiasm and their willingness to pay knee-weakening sums for a scant few bottles. Their reputations seem largely impregnable, but it is not enough to know that the label says *grand cru* because it is the vineyard and not the wine that is ranked as such. Interposed between vineyard and bottle is the hand of the winemaker, clumsy or dextrous, dynamic or coasting, as the case may be.

Such is the dazzle of the famed names that others, particularly to the north near Dijon, are cast into shade. Relatively speaking, this is where to find bargains, for the red wines from north of Gevrey-Chambertin, while not carrying the same depth as their southern neighbours, are still valid expressions of what can be achieved with Pinot Noir in the Côte de Nuits.

MARSANNAY-LA-CÔTE

A strong case could be made for Marsannay's wines being the Côte d'Or's most under-appreciated. History has not been kind, for in times past it was the source of serviceable wine for the nearby city of Dijon but once the railway brought cheaper and stronger wine from the south of France that business withered. After the First World War Joseph Clair-Daü started to vinify rosé from Pinot Noir, an innovation that was an immediate success. Eventually, however, it came to resemble the two-edged sword that is Beaujolais nouveau, in the sense that it established Marsannay's name as a one-trick pony, obscuring all other wines produced there. The lack of its own appellation, rectified in 1987, didn't help either. A move is currently underway to get the best vineyards recognized as *premiers crus*. If it ever happens it will certainly help.

Aside from savvy wine lovers, the public at large has little feel for what Marsannay means, nor any knowledge of how good the best wines now are – they have still to establish an identity for themselves. The rosés and the whites are good but this is red wine country, with echoes of Gevrey's structure in the best wines. Marsannay also has a trump card in the shape of vignerons such as Cyril Audoin, Bruno Clair and Sylvain Pataille. It is their efforts and those of others like them that will see Marsannay better known and better regarded in the future.

Producers

Domaine Charles Audoin

7 rue de la Boulotte, 21160 Marsannay-la-Côte

Tel: +33 3 80 52 34 24

'It is enough,' says Cyril Audoin with a roll of his eyes, as he comments on his family domaine's expansion from an original 2.5 hectares when they bottled their wine for the first time in 1972, through 12 hectares in the 1980s, to 14 today – comprising 11.5 hectares of Pinot Noir, 1.5 of Chardonnay and 1 of Aligoté.

The Aligoté was planted in 1925 and those centenarian vines, wizened by age as they are, produce a wine that bounces with life, fresh and vivacious, with enduring energy on the finish. 'I make it just like my Chardonnay,' says Audoin, further explaining that whole grapes go into the press, followed by stainless steel fermentation and élevage in barrel. The trick is that, while it is made just like Chardonnay, it is not trying to be Chardonnay. In a broader context the wine represents yet more irrefutable testimony to support the case for Aligoté's rehabilitation as a serious grape variety.

Audoin, never lost for a succinct phrase, describes his Marsannay Rosé as, 'a white version of Pinot Noir … and I'd like to do a sparkling version.' It is made from 100 per cent Pinot Noir: 'I treat it like a red in the vineyard and make it like a white in the winery,' says Audoin. After a quick pressing the juice is quite red but ultimately loses colour during élevage to deliver a shimmering rosé of lively red fruit, without the cosmetic overlay that hobbles so many rosés.

Charming though Audoin's Aligoté and Rosé may be, it is upon his well-crafted red wines that his reputation rests. His own favourite is the Marsannay Les Favières, a vineyard with little more than a veneer of soil on top of the limestone, from which he produces a substantial wine, well structured but not weighty. He stresses the need to understand his terroirs and is not afraid to vary his methods to achieve impressive results. When it comes to de-stemming he readily admits that he has no set policy, varying from using 100 per cent whole clusters in 2018, to almost none in subsequent vintages. Stem ripeness is what he seeks: 'Stems help the balance … It depends on the vintage … It's not a recipe … I have to find the best balance.'

Work–life balance is also important for this fanatical rugby fan and as an early September heatwave in 2023 pushed the grapes rapidly towards

68 CÔTE D'OR

maturity, gaining three degrees of alcohol in one week, he worried he would miss the opening game of the Rugby World Cup between France and New Zealand on 8 September. He would have to harvest that day, so he called some friends: 'I need you now!'

On the day he rose at 5.00 a.m.; they picked at 6.00 and pressed at 11.00. He was on the train to Paris at 1.00 p.m., made it to the game and returned home the following morning at 3.00 a.m. France won and the harvest was saved. Audoin found the best balance.

Try this: Marsannay Au Champ Salomon

This vineyard of 5.5 hectares may owe its name to the presence of a gallows there many centuries ago, the dispensers of justice keen to suggest they possessed the wisdom of Solomon. Today, less ghoulishly and in the hands of Cyril Audoin, it produces a round and supple wine, enlivened by a tingle of acid and supported by whispering tannins. There is noticeable intensity and weight but the silky texture and good balance ensure that it doesn't pummel the taste buds.

Domaine Bruno Clair

5 rue du Vieux Collège, 21160 Marsannay-la-Côte
www.brunoclair.com; tel: +33 3 80 52 28 95

As constituted today this domaine dates only from 1979 though its history goes back more than a century to 1919 when, in a matrimonial link that joined opposite ends of the Côte d'Or, Joseph Clair from Santenay married Marguerite Daü from Marsannay. Since then its development has been more than usually complicated; it is traced in detail on the domaine's excellent website. Put simply, disagreements after Joseph's death in 1971 prompted his grandson, Bruno, to strike out on his own and it was only in the mid-1980s that he was able to take control of a good portion of the family vineyards, with some more purchased in the decades since to bring the domaine to its current size of 27 hectares spread across eight villages, producing 32 appellations. About one-third of the vineyards lie in Marsannay, with a sizeable chunk in Gevrey-Chambertin, including a hectare of Clos de Bèze, while the others stretch further south as far as Corton-Charlemagne and Savigny-lès-Beaune.

Today, Bruno and his wife, Isabelle, have been joined by their three children: Edouard in the vineyards since 2010, Margaux in charge of administration and Arthur works in the cellar, both since 2018. They are responsible for a range of wines that compares favourably with all but the most richly endowed domaines in the Côte d'Or: after the hectare of Clos de Bèze come 0.4 in Bonnes Mares and then a generous crop of *premiers crus* in Gevrey-Chambertin, including a hectare in Clos Saint-Jacques and slightly less in Les Cazetiers, followed by many others along the côte. These include two further *premiers crus*: Chambolle-Musigny Les Charmes and Vosne-Romanée Aux Champs Perdrix, with about one hectare in each. The Charmes is everything it should be, silky and smooth, while the Champs Perdrix has sterner tannins in amongst the juicy fruit. 'It's in good form since global warming,' says Arthur Clair of the latter.

Moving about the cellar with Arthur as he draws barrel samples for tasting, the talk is as much about the current vineyard work as the wines themselves. When speaking about replanting a section of vineyard he emphasizes: 'We don't want to make a monoculture again.' Since 2018 some fruit trees – apple and peach, pear and cherry – are now incorporated every nine rows, carefully trained to let the tractors pass.

Cultivation is organic, though not certified, and at harvest the first sorting of the grapes takes place at the vineyard, with a further one at the winery. The grapes arrive in 10–12 kilogram boxes and for those picked in the afternoon a cool room is used to bring the temperature down to 15°C. Between 20 to 50 per cent whole bunches are retained for fermentation in open-topped wooden vats for about 12 days. The aim is for a light extraction with a maximum of six to eight punchdowns, with the temperature not allowed to go above 32°C. Élevage is in 228-litre barrels, up to 50 per cent new, and lasts for 18 to 20 months before bottling.

Together, the new generation at Bruno Clair is bringing a notable dynamism to the domaine and it is not fanciful to suggest it is translating into more conviction in the wines, greater ballast in the flavour as it were. Another decade may yet see those wines accorded a place in the Côte d'Or's pantheon of greats.

Try this: Bonnes Mares grand cru

Bonnes Mares is one of Chambolle-Musigny's two *grands crus* save, that is, for the small portion that lies within Morey-Saint-Denis, all of which is owned by Bruno Clair, giving it some of the status of a *monopole* but without official designation. Arthur Clair describes the wine as, 'the dad of the cellar,' and counsels, 'You have to be careful with extraction,

70 CÔTE D'OR

you must treat it gently or else it is too powerful.' The wine echoes his words: plump, succulent and rich. There is a lavish opulence that could easily run rampant were it not held on a short rein. Treat this to 15 years in the cellar to experience it at its best.

Domaine Sylvain Pataille

6 rue Neuve, 21160 Marsannay-la-Côte

Tel: +33 3 80 51 17 35

The phrase, 'great wine is made in the vineyard,' has been worn to scrap in recent years, yet the kernel of truth that it espouses cannot be denied and if living, walking proof of its veracity is sought, look no further than Sylvain Pataille. He's not hard to spot.

It is sometimes said of winemakers that they produce wines in their own image: the reserved character who produces austere wines, the rotund fellow whose wines are amply flavoured and so on. It is an assertion that could never be levelled at wild-haired and bearded Sylvain Pataille. His wines are models of their kind, beautifully crafted expressions of grape and ground, refracted through the prism of each vintage. The range is wide, almost bewildering, yet each wine carries its own distinctive message – and the carrier in many cases is the Aligoté grape. Indeed, Pataille's cellar might be dubbed 'Aligoté central' such is Sylvain's devotion to this grape.

Sample after sample is poured, each from a specific plot, delivering wave after wave of argument to support Aligoté's claim to potential greatness. A trio provides convincing evidence: Les Auvonnes au Pépé boasts richness and complexity to round out Aligoté's sometimes one-dimensional profile, Clos du Roy takes things a step further by way of expansive fruit and ringing length and Petit Puits, made from 100-year-old vines and producing about one barrel annually, is lavish yet never oppressive thanks to excellent balance: an incredible wine.

The reds, usually made with a high percentage of whole clusters, are wines of remarkable finesse and composure, and all punch well above their nominally humble appellations. La Montagne glides gracefully across the palate, underpinned by strong meaty notes. Le Chapitre is rich and full and, incredibly, was only granted Marsannay appellation status from the 2020 vintage, before which it was a humble Bourgogne. Les Longeroies displays abundant sweet fruit and savoury notes, all corralled by a crisp rasp of tannin.

Pataille cultivates his vines organically, with some biodynamic practices applied too. The resulting wines have an immediate appeal yet these are no easy charmers without staying power, particularly the pure and intense reds. Nor should the excellent rosé be overlooked. This is proper wine, bereft of the saccharine slick that blights some rosés, thanks to a two-year élevage. After a visit to 6 rue Neuve one is left to wonder: what might Sylvain Pataille achieve if granted access to a slice of Chambertin or Chevalier-Montrachet?

Try this: Marsannay L'Ancestrale
This is a true *vieilles vignes*, made from vines up to 80 years of age, located in three different *lieux-dits*: Clos du Roy, En Clémengeot and Les Ouzeloy. Eighty per cent whole bunches are used in the vinification, and élevage lasts a full 24 months. The wine is a slow starter, not giving a lot on the nose and only opening slowly on the palate. Repeated swirling and slurping are needed to coax out the marvellous panoply of flavours. There is delicious, sweet-fruited intensity with a trickle of incense-like spice, all carried by a vivacious lift right into the long finish. If served blind alongside some more patrician names from nearby Gevrey-Chambertin it would not be shamed – and might well confound firmly held opinions when its identity is revealed.

FIXIN

It might not be fair to label Fixin as a poor man's Gevrey but it is not too far wide of the mark and is not entirely dismissive to say so. There's a touch of praise there also, an acknowledgement that the wines may be less noble, but that they also share some of the same DNA without the crushing prices that Gevrey's best command. It is nearly forty years since I was first advised to seek out good Fixin if funds wouldn't stretch to Gevrey, and it is advice that still holds good today.

The commune of Fixin includes the hamlet of Fixey and the vineyards reach up the slope to altitudes of 350 metres where the *premiers crus* are located. The adjacent forest houses the Parc Noisot, named after Claude Noisot who was a member of Napoleon's imperial guard and who so revered the emperor that he commissioned the *Réveil de Napoléon* statue in the grounds that depicts Napoleon awakening, Lazarus-like, to face the rising sun.

72 CÔTE D'OR

Producer

Domaine Pierre Gelin

22 rue de la Croix Blanche, 21220 Fixin

www.domaine-pierregelin.fr; tel+ 33 3 80 52 45 24

This domaine celebrated its centenary in 2025, having been founded by the eponymous Pierre Gelin in 1925. Its modern incarnation, however, dates from 2010 when a brand new winery and underground cellar was inaugurated. Prior to this, winemaking operations were spread across two wineries, two barrel cellars and seven other cellars. The logistics must have been bewildering, though in conversation Pierre-Emmanuel Gelin, son of Stéphen and grandson of Pierre, noted that the malos were easier to manage in the fragmented past.

The domaine runs to about 13 hectares with the majority in Fixin and the balance in Gevrey-Chambertin. There are two *monopole* vineyards – Clos Napoléon in Fixin and Clos de Meixvelle in Gevrey – and these represent the pillars upon which the domaine's reputation rests. The former, which lies above the village and near the forest, at about 300 metres elevation, was purchased by Pierre Gelin in the 1950s but only half a hectare was planted to vines then. There were also fruit trees, and after these had been uprooted Gelin set about creating three terraces and planting them with vines. The effort was worth it, the wine majors on confit red fruit and spice with a rustic core of tannin adding structure and length.

The label carries a reproduction of the *Réveil de Napoléon*, which gives it immediate visibility. Indeed, in an almost unique practice, each of the Gelin wines carries its own individual label, though none are quite so eye-catching as the Clos Napoléon. The Clos de Meixvelle (also spelt 'Mévelle') makes do with a more conventional iteration. The wine itself comes from a small, 1.8-hectare vineyard that lies in the heart of the village of Gevrey and the *clos* is surrounded by 4-metre-high walls that create a microclimate, with further character being added by the shallow soil. There is impressive depth and intensity of fruit, rather than the more classic muscle of Gevrey.

The range also includes a Bourgogne Passetoutgrains, an Aligoté and several Fixins, all topped off by the Chambertin-Clos de Bèze (below). Organic practices were introduced in 2015, with certification achieved in 2019, and all the wines see an extended élevage of at least 20 months in barrel before bottling. Then, in another slightly unusual practice, they are cellared at the domaine for a year or two before release.

Try this: Chambertin-Clos de Bèze grand cru

The domaine's holding amounts to 0.6 hectares comprising two plots in the southern half of the vineyard, close to the *Route des Grands Crus*. This jewel in the crown was bought by Pierre Gelin in the 1950s, prior to which he rented the land. Élevage in barrel, about 50 per cent new, lasts for almost two years. The result is a wine of substance and structure, full-fruited but not too weighty thanks to good balance. Mineral and herbaceous elements combine to add complexity on the persistent finish.

GEVREY-CHAMBERTIN

For some, particularly traditionalists who have yet to acknowledge the quality wines coming out of Marsannay and Fixin today, the Côte d'Or truly begins at Gevrey-Chambertin, where starts the catalogue of storied *grands crus* vineyards. Gevrey vies with Vosne-Romanée as the Côte de Nuits' most celebrated commune. In purely vinous terms, Vosne shades it thanks to an allure and renown unequalled in the wine world, but the village itself has a buttoned-up feel compared to the relative bustle of Gevrey, which presents the same prosperous face to the world as Meursault in the Côte de Beaune. Gevrey doesn't hide its light. Nor has it ever; it is at Gevrey's door that blame must be laid for the hyphenated complication found in the village names of the Côte d'Or, for it was Gevrey that first appended the name of its most famous vineyard in 1847 to emerge as newly minted Gevrey-Chambertin. Wine branding and image creation is not a new thing.

The village itself sprawls somewhat, interlaced with vineyards, and drifts downslope from its famed *premier cru* Clos Saint-Jacques towards and across the main road, tracking a similar drift in the vineyards all the way. Compared to many of the côte's smaller villages there's heft and substance in Gevrey, just as there is in the wine. It's a village worth exploring on foot, though it is hard to pinpoint a defined centre; you always seem to be approaching it without actually arriving.

Traditionally, the wines of Gevrey-Chambertin made strong statements; complexity came by way of a panoply of bold flavours rather than nuanced layers of sophistication. They were solid and satisfying with

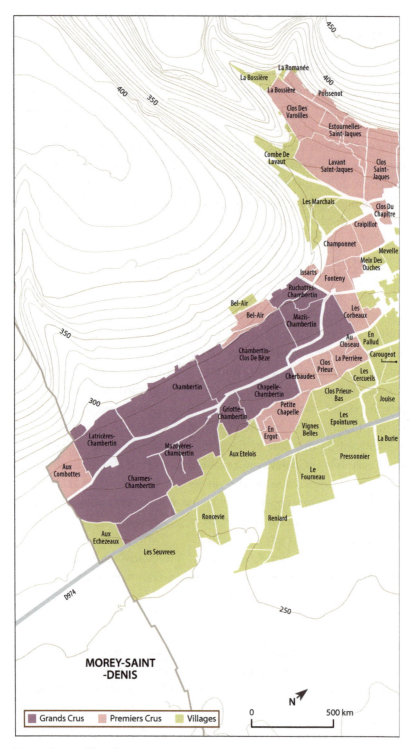

Map 2: Gevrey-Chambertin

THE VILLAGES AND PRODUCERS OF THE CÔTE DE NUITS 75

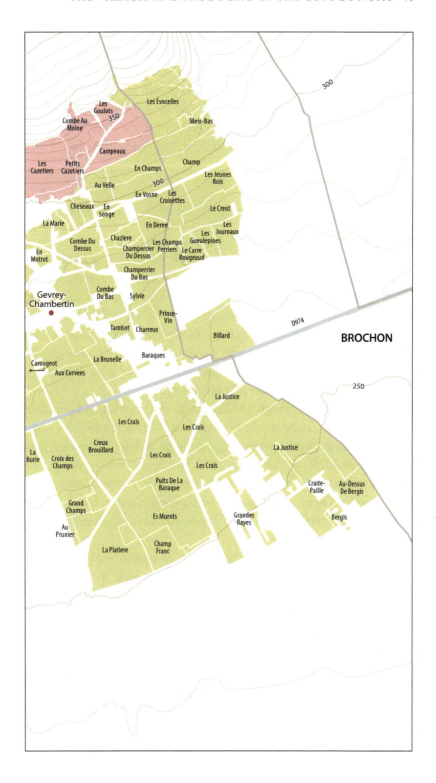

76 CÔTE D'OR

tremendous depth, wines of concentration and power, defined by robust flavours, often rough-edged in youth but eventually giving way to satin texture and engaging harmony when mature. Today, the flavour profile has lightened, the meat and muscle of old is still found but in a more refined package; there is more harmony and less heft. That said, Gevrey remains the most ageworthy, and 'age-needy', of all the Côte d'Or's reds, when robust youth cedes to satin texture and engaging maturity.

Producers

Domaine Les Astrelles

1 route de Saint-Philibert, 21220 Gevrey-Chambertin
www.domainelesastrelles.fr; tel: +33 3 80 33 46 03

The genesis of this domaine dates to 2010 when husband and wife, Jean-Marie and Isabelle Chapier, both from medical backgrounds, began making wine in Bergerac. Nearly a decade later Burgundy called and in 2019 they bought Domaine Pierre Naigeon in Gevrey-Chambertin, installing Julien Millet as cellarmaster the following year. Chapier describes himself as, 'an investor who wants to get involved,' though not to the point of micromanaging his team, who are able to go about their business without him permanently at their shoulders.

Millet is the son of François Millet (see p. 105) and brings boundless energy to his role, as he climbs between stacks of barrels to draw samples, elaborating on his winemaking as he goes. Fifty different wines were made in 2023, up from 39 in 2022, which makes for a convoluted trek about the cellar to taste a selection of them. It is a bewildering array and besides calling for a cellarmaster with a good memory it may be difficult for consumers to grasp. Back labels with accurate details of each wine's provenance would help.

'Whole bunch can obscure the reading of terroir,' says Millet, explaining that all the fruit is 100 per cent de-stemmed, 'to preserve the character of each vintage'. Up to and including the 2023 vintage, fermentation was in stainless steel and concrete but since the inauguration of a completely new winery in 2024 it is all in stainless steel. There is no cold soak and only pumping over is used during the 15 to 25 day vinification. 'It's an infusion style,' says Millet, 'we have to be cautious with the ripeness of the skins.' A gentle pressing follows in a pneumatic press, 'we lose some juice but keep the finesse,' though he hopes in future to change to vertical pressing: 'I love it, the cake acts as a filter.' Then to barrel for 15 months and a month in stainless steel before bottling.

At present, production favours red wines, 70:30, but Millet is planning to increase the proportion of white to follow consumer trend. The white grapes go straight to the press and the juice is allowed to settle for one to two days before fermentation in barrel. A wide variety of barrels is used for both colours, such as Radoux, Tonnellerie de Marsannay and Tonnellerie Berger but also includes the currently fashionable Stockinger from Austria and the extremely limited production from Eric Millard.

Domaine Les Astrelles comprises a dozen hectares, with a further six of bought-in grapes, and produces wines from Puligny-Montrachet to Marsannay, spanning almost the length of the Côte d'Or. The Pommard 1er cru Les Arvelets has energy aplenty but the tannins are silky and not aggressive, 'quite elegant for a Pommard,' says Millet. The Nuits-Saint-Georges 1er cru Aux Murgers is similarly refined, expanding in the mouth with lovely fruit and a gentle undertow of tannin, 'a Vosne in disguise'. There is also a tiny production of Mazis-Chambertin, about 200 bottles, from the top of the vineyard where there is virtually no soil and the vines grow on the rock.

'Astrelles' is a portmanteau name – a combination of 'Astro' and 'Parcelles' – and works well in French and English, which is a small but crucial consideration for a new domaine, seeking to establish a place in the market.

Try this: Gevrey-Chambertin 1er cru Lavaux Saint-Jacques

Lavaux (or Lavaut) is next-door neighbour to the celebrated Clos Saint-Jacques, and the Astrelles holding there is a minute 0.08 hectares that faces south-east, yielding 400 bottles a year on average. Vine age is about 50 years. Julien Millet declares himself, 'very glad to make some Lavaux,' going on to describe their plot as, 'a tiny *grand cru*'. The immediate impression on the palate is of sweet fruit that toys with lushness but which avoids that trap thanks to deeper, sterner notes that form the foundation of the wine. 'We have the power but the right power, there isn't too much,' Millet adds. The fruit and the power carry on in harmony to a lengthy, satisfying finish.

Domaine Philippe Charlopin

18 route de Dijon, 21220 Gevrey-Chambertin
www.domaine-charlopin-parizot.com; tel. +33 6 24 71 12 05

The stereotypical image of a Burgundian family domaine centres on fusty cellars bedecked with cobwebs and mould, where a ball of twine might

78 CÔTE D'OR

prove useful for navigation. Not so at Domaine Philippe Charlopin, where the cellar has something of the look and feel of a Bordeaux *chai*. Ranks of barrels stretch into the distance, framed by a lattice of pillars and graceful arches. It's well lit, there isn't a cobweb in sight and the smooth, even floor allows Charlopin to wheel a spittoon about easily, as he searches out specific barrels for tasting. The contrast between pristine barrel hall and his nearby office, stacked high with the paperwork that is the bane of every vigneron's life, is marked.

Philippe Charlopin started with a mere 1.5 hectares of vineyard in 1976 and by 2006 had expanded the holding to such an extent that a move to large, new premises on the main road between Gevrey and Brochon was necessary. Today, assisted by his son Yann, he makes three dozen wines from 25 hectares, including an enviable roster of *grands crus* headed up by 0.2 hectares of Chambertin, in a strip towards the south of the vineyard that runs almost its full width.

To experience the quality of Charlopin's wines, however, it is not necessary to drink at *grand cru* level. The Marsannay En Montchenevoy is a firm and fruity delight, replete with cherries, raspberries and a ping of acidity. While tasting this with Philippe Charlopin I ask if Marsannay will ever achieve *premier cru* status for its best vineyards. The answer is non-verbal – a classic Gallic shrug and then a grin – but with wine of this quality it hardly needs it.

Satisfying though the En Montchenevoy is, it is understandably outstripped by the Gevrey-Chambertin Vieilles Vignes. This is a complete wine in every sense, assembled from plots of old vines in the north, centre and south of the commune. Charlopin describes it as *puissant*, and it is powerful, but in a restrained and focused idiom that rewards repeated sipping. Further up the pecking order, the Mazis-Chambertin, from a holding of less than one-tenth of a hectare in the upper part of the vineyard, adjacent to Ruchottes-Chambertin, is an extraordinary wine. To begin with it is completely closed in on itself and needs repeated swirling to coax the flavour out, finally delivering a cornucopia of scents and sensations, where bountiful fruit is leavened by earth and spice, with meatier notes in the background to add savour.

All the Charlopin wines display substance and structure but none is overbearing; there is balance and refinement too. They are usually 100 per cent de-stemmed but in richer vintages some whole clusters will be used. Élevage is in a mixture of barrels from five coopers: Rousseau, Chassin, Billon, Seguin Moreau and Atelier Centre France. For the white wines 500- and 600-litre demi-muids are used. But it is on the reds that Charlopin has built his reputation, which he continues to burnish with wines of exuberance and concentration, vintage after vintage.

Try this: Vosne-Romanée 1er cru En Orveaux
Much of the Côte d'Or's complicated nomenclature is summed up in the name of this wine. 'En Orveaux' is a *lieu-dit* of Flagey-Echézeaux (which can be labelled as Vosne-Romanée), the upper part of which is ranked *premier cru*. The lower part lies within the *grand cru* Echézeaux – and Charlopin also owns vines in that section of Echézeaux. This wine, however, is simple in the best sense. The flavour is pure and penetrating, linear and long. It is so seamlessly woven together that trying to identify individual components is a wasted exercise.

Domaine Claude Dugat

1 place de la Cure, 21220 Gevrey-Chambertin
Tel: +33 3 80 34 36 18

'I see you are building a new house,' I said to Claude Dugat, when leaving his domaine after my first visit in 2007. 'That's a stable for my horse,' he replied. That horse now ploughs the great vineyard in the sky and has not been replaced, the ploughing of the *grands crus* is now done by contract.

Claude Dugat has retired and the domaine is in the hands of his three children, Bertrand, Laetitia and Jeanne. In training them to succeed him he was careful to ensure that each learned all the operations of the domaine so that today, as Laetitia puts it, 'we make collegial decisions,' further explaining that, because they can all multi-task, they are not bound to the domaine during the summer when ripening needs to be monitored. Holidays can be taken in rotation.

The domaine may be small but its reputation is stellar. In Burgundian terms its history is not long, being founded by the siblings' grandfather, Maurice, when he bought the current premises in 1955. Maurice was a *Cadet de Bourgogne*, one of the jolly band of singers who provide entertainment at the *chapitres* of the *Confrérie des Chevaliers du Tastevin*. As such, he travelled the world spreading the Burgundy message, 'from Tokyo to Washington,' Laetitia explains, and although he lived to be 92, dying in 2002, 'he never realized how successful we had become.'

The new generation has not made any radical changes in style, instead gently ushering in a softening in the sometimes massive flavours

80 CÔTE D'OR

of yore. (The Gevrey-Chambertin 2004, 2005 and 2006 remain tight-coiled and uncommunicative at time of writing.) There is more elegance to balance the structure, seen to good effect in the Gevrey-Chambertin 1er cru, blended from two vineyards: Craipillot, which looks across the road to Clos Saint-Jacques; and La Perrière, across the road from Mazis-Chambertin. It strikes an appealing balance between the elegance of today and the reserve of yesterday, without any conflict on the palate. Moving up the scale, the Lavaux Saint-Jacques 1er cru has some of Gevrey's signature meaty character but there is clarity of fruit too, with a smooth texture supported by firm but civil tannins. 'We like vivacity and freshness,' says Laetitia, going on to explain that the Dugat policy is to harvest relatively early: 'the picking date is critical'.

Maison La Gibryotte is an associated small négociant that helps to augment the domaine's production and, with an eye to the future, will hopefully maintain production on a scale sufficient to provide for the three siblings. Domaine Dugat is on song right now; it would be a shame to see it divided amongst succeeding generations, as happens so often in the Côte d'Or.

Try this: Charmes-Chambertin grand cru

Charmes-Chambertin lies directly across the *Route des Grands Crus* from Chambertin itself, yet while the latter expresses itself with power, Charmes presents a more delicate countenance. Living up to its name, you might think, but 'Charmes' in this instance derives from 'chaumes', meaning lands that are difficult to cultivate, suitable only for vines. Regardless, the Dugat version has charm aplenty, starting with a satin texture that opens slowly to reveal a firm structure and great depth. Balance and harmony lead gently into a long, satisfying finish.

Domaine Duroché

48 rue de l'Eglise, 21220 Gevrey-Chambertin
www.domaine-duroche.com; tel. +33 6 08 37 33 49

Some years ago, when shopping in Chassagne-Montrachet's Caveau Municipale, I enquired about Gevrey-Chambertin, emphasizing that I was looking for a true expression of that commune's style. The assistant immediately pointed to Duroché, a then seldom-heard-of name, dubbing him the 'rising star' of the appellation. Pierre Duroché's star is now well and truly risen and there can be no doubt about the quality of his wines, which I buy without hesitation.

THE VILLAGES AND PRODUCERS OF THE CÔTE DE NUITS 81

Duroché is softly spoken but articulate and, together with his wife Marianne (with whom he has a small négociant operation), he brings a considered, thoughtful approach to bear on his winemaking: 'Each vintage we have more experience. Marianne and I both have the same idea of the wines we want to make … We want to improve the details … I think the wines are more precise now.' Not that they were imprecise previously. As I wrote of his Chambertin-Clos de Bèze in the first edition of this book: 'The wine is rounded and complete, delicate but not frail, pure and precise.'

In addition to their precision, the wines have greater assurance now, brought about by Duroché understanding, and never underestimating, the challenge of achieving the style he seeks: 'It is always difficult to have something very dense but also elegant at the same time.' That, in a nutshell, describes the essence of his wines. There is no lack of structure and substance but equally impressively there is also harmony and elegance.

Many organic practices are followed in the vineyards, without certification, and at the winery the grapes are largely de-stemmed, save for 10 to 15 per cent in the top wines: 'Whole bunches add some complexity but we don't want stemmy green flavours,' says Duroché. The berries are cooled for two or three days and fermentation takes place in stainless steel, with the cap barely pushed down, simply to keep it wet early on, and some pumping over later. One year in barrel follows with no racking and no new oak for the *grands crus* because of the small quantities: 'I prefer to keep the taste of the fruit,' he says. The wine spends two months in tank before bottling with natural corks – pushed slightly below the rim of the bottle since 2022, so that a blob of wax can be added as a sealer, obviating the need for a 'not environmentally friendly' capsule.

Duroché is master of the micro cuvée, with parcels of 0.02, 0.07 and 0.13 hectares in, respectively, Griotte-Chambertin *grand cru* and the two Gevrey *premiers crus*, Cazetiers and Estournelles Saint-Jacques. The Griotte is the smallest *grand cru* holding in the Côte d'Or, smaller even than Taupenot-Merme's 0.04-hectare parcel in Clos des Lambrays. From it, Duroché might make 100 bottles of wine.

There is also a single white wine, a Bourgogne Chardonnay of real class, taut and incisive with great intensity and length, and well worth following. There is a small cohort of domaines where the vigneron's name on the label can be taken as an automatic guarantee of quality. Duroché is one.

82 CÔTE D'OR

Try this: Gevrey-Chambertin 1er cru Lavaut Saint-Jacques Vieilles Vignes

Compared to some of the tiny parcels Pierre Duroché owns, his holding in Lavaut Saint-Jacques feels generous, at 1.2 hectares, which allows him to separate out this old vine cuvée. He produces a barrel, 'or maybe two,' every vintage from the centenarian vines planted in 1923. Inky rich fruit on the nose, with head-filling scents and aromas, presages an exotic palate suffused with sweet incense and dark liquorice. The texture is sumptuous; the flavour is ripe and rich, the wine is dense and long, yet somehow not heavy.

Domaine Sylvie Esmonin

1 rue Neuve/Clos Saint-Jacques, 21220 Gevrey-Chambertin

Tel: +33 3 80 34 36 44

The address says it all, for Domaine Sylvie Esmonin can truthfully claim to be, not just beside, but in the famed Clos Saint-Jacques vineyard. Vines lap right up against the buildings and wrap around to one side – as back gardens go, the *grand-cru*-in-all-but-name takes some beating.

A tad over 6 hectares in Gevrey-Chambertin, spread across 15 plots, make up the bulk of this domaine. Two cuvées of Gevrey are made, a *village* wine and a Vieilles Vignes. The *village* sees about 10 per cent new oak and majors on smooth fruit wrapped around a stern, spicy core. It also ages superbly; the challenging 2016 vintage, where the yield barely reached 20 hectolitres per hectare, was magnificent at eight years old, thanks to a delicious rush of fruit framed by supple tannins. The Vieilles Vignes is a selection from the best parcels of old vines, with an average age of 60 to 70 years and some as old as 100. On entry the texture is smooth and the flavour deceptively light but then the wine unfolds to reveal ample fruit with smoky, meaty notes adding complexity to the persistent finish.

At a humbler level the Bourgogne Rouge, typically made with 50 per cent whole bunches in a warm vintage but completely de-stemmed in a bad year, is stern rather than charming. No new oak is used and there is a forthright juicy rasp on the palate that suggests three to five years of ageing to add harmony.

Significant percentages of whole bunches are used across the range, particularly in the more prestigious wines but, as at all quality-conscious domaines, there is no written-in-stone diktat to be followed slavishly. Fermentation is in stainless steel and wooden vats, the latter for the top

THE VILLAGES AND PRODUCERS OF THE CÔTE DE NUITS 83

wines, with some pumping over for about 10 days, followed by gentle punching down. For élevage the wines spend a year in Dominique Laurent barrels and then six months in tank before bottling.

Try this: Gevrey-Chambertin 1er cru Clos Saint-Jacques

Herein lies a tale. In précis: Sylvie Esmonin's grandfather worked for the Comte de Moucheron, at one time sole owner of Clos Saint-Jacques. When the comte sold up, Esmonin bought a slice and his house, which is where the domaine is based today. Whatever sacrifices *grand-père* had to make to afford the purchase they were well worth it. This is a full-throttle wine, inviting descriptors such as 'lavish', 'ample' and 'robust'. Generous fruit and rich tannins, tobacco and spice, coffee and chocolate all cavort across the palate. It is plump and full yet somehow does not overwhelm. Age brings greater composure as the elements settle and integrate to deliver a wine of impressive fortitude and singing length.

Domaine Fourrier

7 route de Dijon, 21220 Gevrey-Chambertin
Tel: +33 3 80 34 33 99

Words like 'gentle', 'delicate' and 'refined' crop up repeatedly in my tasting notes for the wines of Jean-Marie Fourrier. 'Frail' never makes an appearance. That is the essence of Fourrier's wines, they sing pure and true, relying on finely etched flavours, rather than over-concentration, to make an impression. They are as much about texture as taste.

Fourrier took over this 10-hectare domaine from his father in 1994 and quickly set about putting his own stamp on how things were done. A sensitive hand employed at every stage of the winemaking process, from vineyard to bottle, best describes his approach. Thus the house style sits at variance with the Gevrey-Chambertin stereotype, where structure and robust flavours play a central role.

Despite having a 'lighter weave' as it were, the wines lack for nothing in energy and intensity and it takes only a sip of the Bourgogne Rouge to discover that an absence of overt flavours clears the way for finesse and subtlety to shine. This wine is everything that it should be and in some respects it can be seen as the domaine's standard bearer, for it serves as a liquid mission statement, thanks to an electric charge of flavour driven by crisp fruit and tingling acidity. Made from bought-in grapes, sourced across the Côte de Nuits, thirty thousand bottles were produced in 2022. The wine also flies the flag for Fourrier's firm belief that there should be an accessible entry-level wine, particularly for consumers new to Burgundy, now more so than ever when there is a risk they will be put off by the headline prices paid for the *grands crus* and such like.

At a more exalted level the Gevrey-Chambertin Vieille Vigne comes from 16 different parcels with an average vine age of 60 years, the oldest of which were planted in the period between 1928 and 1945. It is fresh-scented, intense and juicy, while the Gevrey-Chambertin 1er cru Cherbaudes, northern neighbour of Chapelle-Chambertin, is fuller but still cleaves to the precision-flavoured house style thanks to splendid, though never overbearing, intensity. The style may not satisfy those who hanker after old-school Gevrey but any who seek thrill on the palate need look no further.

Try this: Gevrey-Chambertin 1er cru Combe aux Moines Vieille Vigne
This vineyard sits at variance with the mind's-eye image of a Côte d'Or *premier cru*. There is a wild, untamed feel, given emphasis by the vertical back wall of rock from which heat is reflected onto the vines – which are overlooked by an elevated statue of a hooded monk. The wine is a child of its origin, a slight outlier in the Fourrier stable, in the sense of fuller flavours and greater density, with the tannins more forward than its stablemates. But Fourrier's hand keeps them civil when they might otherwise obtrude. Two bottles of the 2007, drunk a month apart in 2024, displayed rather different characters, the first still marked by tannin, the second more forthcoming. A gorgeous combination of fruit and spice, with the tannins in support rather than dominating. (And yes, the 'Vieille Vigne' is singular, not plural.)

Domaine Henri Rebourseau

10 place du Monument, 21220 Gevrey-Chambertin
www.rebourseau.com; tel: +33 3 80 51 88 94

In vinous terms a yawning chasm separates Burgundy's Gevrey-Chambertin and Bordeaux's Saint-Estèphe, though the two share similar reputations as the four-square faces of their respective regions. However, a stronger link has been in place since 2018, when the Bouygues family, owners of Château Montrose, bought a majority stake in Domaine Rebourseau and commenced a massive programme of construction and renovation to deliver a magnificent new winery in time for the 2023 vintage. Any who have visited the jaw-dropping barrel hall at Montrose will not be surprised to learn that the new facilities at Rebourseau are similarly impressive, if a little less grand.

It was a bottle of Gevrey-Chambertin 2019 that alerted me to the new broom sweeping away the cobwebs at this historic Gevrey domaine, which dates from 1782. Though not especially large – 13 hectares – the *grands crus* holdings are eye-watering: well over a hectare in Charmes-Chambertin, almost a hectare in Mazy-Chambertin, about half that in Chambertin, one-third in Clos de Bèze and over two hectares in the heart of Clos de Vougeot.

Bastien Giraud has been in place as manager since 2022, having previously worked at Faiveley and, while standing at the grape reception area, he repeatedly emphasizes the need for rigorous sorting: 'Sorting is the key,' he says, explaining that the grapes are first sorted at the vineyard, then on a vibrating table and finally on a conveyor-belt table, manned by eight people. 'Each cluster must be touched,' he says, turned over and examined for any sub-standard berries. Thereafter, to de-stem or not is currently the big question at Rebourseau. Some whole clusters are being used in the Charmes-Chambertin and the Chambertin: 'We will proceed with care … It brings complexity … When it is managed properly it is amazing.' Vinification follows in open-top wooden vats from Taransaud, with punching down early on and no pumping over, finishing with a few days' post-fermentation maceration. Élevage is almost exclusively in 228-litre *pièces*, 40 to 50 per cent new for the *grands crus*, 30 per cent for the other wines.

The results are compelling expressions of all that is good about Gevrey-Chambertin: wines replete with dark fruit and some heft, shot through with delicious grace and energy. It is a beguiling amalgam, seen initially in the Gevrey-Chambertin La Brunelle and, most convincingly, in the Chambertin, which moves ahead of its *grands crus* siblings with greater volume rather than outright weight in the mouth, to deliver a ringing endorsement of the new regime.

Note: as this book went to press it was announced that Sylvain Pabion had succeeded Bastien Giraud as manager at Rebourseau.

Try this: Mazy-Chambertin grand cru

Mazy (also Mazis) is divided by a winding road into upper, *haut*, and lower, *bas*, sections covering 9 hectares in total, and Rebourseau is the third largest owner, the holding being spread across three parcels in the lower section. This is a beautifully composed wine, quiet on the palate at first, before the flavour unwinds and develops. Dark fruit, suffused with meaty, liquorice notes dazzles the taste buds and the excitement continues long into the exuberant finish. For Bastien Giraud: 'It is the type of *grand vin* I really appreciate; there is aromatic power and concentration and it is really well balanced.'

Domaine Rossignol-Trapet

4 rue de la Petite Issue, 21220 Gevrey-Chambertin

www.rossignol-trapet.com; tel: +33 3 80 51 87 26

When Nicolas Rossignol compares the Rossignol-Trapet style to the Gevrey-Chambertin stereotype he describes it thus: 'It is lighter with more elegance. Sometimes less is better, we do less pressing and no *pigeage*, we like to keep the elegance of Pinot Noir. We are not worried about the colours, we try to respect the terroir, we like to have stems.' He stops short of saying, 'and the results are marvellous,' leaving hyperbole for others.

The wines are excellent and manage to pull off the almost contradictory challenge of combining grace with gravitas. Thus, if it is old-school Gevrey that you are after – wines replete with commanding, sometimes forbidding, flavours – you need to look elsewhere. The 'meat and muscle' Gevrey stereotype holds no sway here. Instead, the palate will be rewarded and charmed by precision flavours, beautifully crafted. 'Intense not concentrated,' might be their motto.

Nicolas Rossignol and his brother, David, run the domaine on biodynamic principles, yet there is no rigidity in their winemaking philosophy. De-stemming decisions are flexible, Nicolas explaining that, 'it depends on the typicity of the grapes before harvest'. Thus, in three successive vintages, 2021 saw one-third whole bunches, rising to twothirds in 2022 and back to half in 2023. Generally, more whole clusters are used in the *premiers* and *grands crus* than in the *village* wines. Gentle winemaking, 'we adapt the vinification to the grapes' origin,' then follows, with some pumping over, before élevage in Rousseau and Chassin barrels: between zero and 15 per cent new for the basic wines, 25 per cent for the *premiers crus* and about 45 per cent for the *grands crus*.

Unlike most domaines, whose vineyard holdings are relatively generous at *village* level, rising to minuscule parcels of *grands crus*, if at all, Rossignol-Trapet is well endowed at both ends of the spectrum. There are 5 hectares of *village* Gevrey, scattered across several plots that are vinified separately and then amalgamated into a sweet, soft wine of discreet structure. At the top, the Chambertin (below) is the flag carrier but its shadow should not obscure the Latricières-Chambertin (0.73 hectares) and the Chapelle-Chambertin (0.54 hectares). The former has much the same soil as Chambertin but is a little cooler thanks to some breeze from the nearby Combe Grisard, yielding a dark-fruited wine with a refined tannic structure; the latter has greater amplitude with a suave texture and impressive length. They might have to bend the knee to the Chambertin but only just.

Try this: Chambertin grand cru

Rossignol-Trapet is one of the largest owners of this regal *grand cru*, laying claim to a generous slice of 1.6 hectares that runs almost the full width of the vineyard, from the road nearly to the tree line, rising about 15 metres in the process. This wine cleaves to the house style – power and weight sing in the chorus here, not solo. The overall impression is as much about the svelte texture as the parade of sweet, dark fruit flavours that are leavened by a wispy note of liquorice, before departing into a pure, long finish. Tasting the 2022 from barrel my note concluded, 'No words …'.

Domaine Sérafin Père et Fils

7 place du Château, 21220 Gevrey-Chambertin
Tel. +33 3 80 34 35 40

'I no longer take my holidays in August,' says Frédérique Goulley, niece of Christian Sérafin, who has overseen winemaking here since 1999. Such a remark outside of winemaking circles might simply chivvy the conversation along, but when spoken by a Côte d'Or winemaker it assumes greater significance, delivering a message that climate change is real and unavoidable. More scientific measures may confirm its influence but this simple statement is both forceful and memorable. The winemakers' annual *vacance*, once a lapidary fixture in August, has shifted.

Goulley is not short on memorable quotes: 'I am like Obelix, I fell into the wine barrel when your father christened me with Charmes-Chambertin,' this latter, metaphorical, remark is addressed to her cousin Karine, who has worked at the domaine since 2007. In the vineyards they are ploughing less than previously to protect the soil and following biodynamic practices, though without certification because of 'too much administration'.

In the winery, after a relatively early harvest, the grapes are fully de-stemmed when the berries are mature though the stems are not. Five days of cool maceration at 12°C follows, before fermentation in stainless steel with a combination of *remontage* and *pigeage*. Thereafter, élevage is in all-new barrels from François Frères, Chassin and Seguin Moreau, Goulley liking the 'regular' tannic influence she gets from the wood. Structure and substance abound in the finished wines yet they carry their weight well and don't belabour the palate with oppressive flavours.

Outside of 'home base' in Gevrey-Chambertin the holdings include one-third of a hectare each of Chambolle-Musigny *premier cru* Les Baudes and Morey-Saint-Denis *premier cru* Les Millandes, the latter purchased in 1996 after being advertised in the newspaper, an unimaginable necessity today. The Chambolle is a gorgeous wine full of energy and tingling intensity, while the Morey is less memorable, sound rather than special.

Meanwhile, the Gevrey-Chambertin Vieilles Vignes is an exemplar, both for the commune and the domaine, full and earthy at its core but with a smooth, textural overlay courtesy of deft winemaking. Further

Domaine Sérafin cellars, Gevrey-Chambertin

THE VILLAGES AND PRODUCERS OF THE CÔTE DE NUITS 89

up the ranks the Gevrey-Chambertin 1er cru Les Corbeaux comes from a walled, half-hectare *clos* within the vineyard, the walls protecting it from spring frost in years like 2016. The wine is vividly flavoured, deep and long, with the oak remarkably well integrated even when young. The Charmes-Chambertin *grand cru* (which includes some Mazoyères) takes things a step further: concentrated and firm and only missing a decade or two in the cellar.

Domaine Sérafin spans a modest 5.5 hectares, yet even at that level of production the house policy is to hold back some stock as insurance against spring frost and summer hail, selling two-thirds of each vintage once bottled and the remaining third after five or six years. That remaining third should have great appeal.

Try this: Gevrey-Chambertin 1er cru Les Cazetiers

Cazetiers is a 9-hectare vineyard that faces almost due east with only a slight turn south and it can be accessed by stepping no more than a couple of paces from the Sérafin cellar. They own about one-third of a hectare, planted in the 1960s, the remainder being divided between a dozen or so others. It is a cooler terroir and is the last of the domaine's vineyards to be harvested. The nose says it all, a perfumed aroma of dark fruit that presages an equally compelling palate; there is weight and structure but the tannins support the fruit without obtruding, while the acidity delivers a fresh note to keep the finish clean.

Domaine Trapet Père et Fils

53 route de Beaune, 21220 Gevrey-Chambertin
www.domaine-trapet.fr; tel: +33 3 80 34 30 40

'It depends on what we see, we are not dogmatic. You have to be very confident in your grapes. In 2021 we de-stemmed 100 per cent, in 2022 it was 50 to 80.' Jean-Louis Trapet is speaking inside what could be described as a 'tasting tunnel' created from hundreds of barrel staves fixed together to create an elongated arch that leads from his office to the winery. 'Tasting room' is too prosaic a term.

One senses he would prefer to talk history than winemaking and his comments regularly veer that way, stretching back to the nineteenth century and an ancestor's battle with phylloxera. Family history took concrete shape later when, in 1904, Louis Trapet bought his first vineyard, a parcel in Latricières-Chambertin. Being custodian of such a legacy is quite the responsibility but at Domaine Trapet history inspires

90   CÔTE D'OR

rather than inhibits, stimulating innovation not stifling it. Nothing is fossilized or set in aspic – witness the brightly coloured cartons and cardboard cases the wines are packed in.

More pertinently, Trapet's embrace of organic and biodynamic practices dates to a time when it was more likely to be met with a raised eyebrow than a nod of approval. Such an approach chimes with today's zeitgeist but it was pioneering back in 1995. The results are made manifest in wines such as the Gevrey-Chambertin, where Trapet owns 4.5 hectares scattered across 10 parcels. It is a wine of real breed, gently robust, well-structured with no hard edges. After that, various Gevrey *premiers crus* jostle for attention, notably Les Corbeaux. 'It's not crows!' says Trapet, explaining that the word derives from an old French adjective, *corbe* meaning concave, referring to a slight declivity in the vineyard and not the contemporary word for crow. Complications of toponymy apart, this is a wine of style and opulence, with abundant floral, fruit and spice flavours.

The Trapet Chambertin stands apart. Made from three parcels amounting to almost two hectares, it is a big and forceful wine, though not aggressive, with a near-fathomless depth of flavour. 'One hears the clang of amour in its depths,' opined Irish wine writer Maurice Healy nearly 90 years ago when speaking of Chambertin – the Trapet version chimes perfectly with his memorable words.

From outside, Domaine Trapet presents an unremarkable face to the world and one could easily drive past without noticing – were it not for a giant arrow on an opposite gable wall pointing to it. Any fuddy-duddy impressions the bland exterior might convey are swept away by a walk through the winery. There, amongst the expected oak barrels and vats, are found stainless steel barrels, concrete eggs, glass globes and stainless steel eggs. You will see them all at this address – as well as a massive wall mural depicting the family at work in the vineyards.

Try this: Bourgogne Passetoutgrains 'A Minima'
This 50:50 blend of Pinot Noir and Gamay may be the humblest wine in the Trapet line-up but in some respects it is the standard-bearer for the domaine's low-intervention philosophy, having been made without sulphur since 1996. Suffused with bountiful red fruit, majoring on raspberries, cherries and strawberries, it bounces across the palate, lively and vital, each sip suggesting another. It is the sort of wine that, in all

THE VILLAGES AND PRODUCERS OF THE CÔTE DE NUITS 91

the noise and bother about celebrated *grands crus*, tends to get over-looked. It deserves not to be.

MOREY-SAINT-DENIS

It's common to say that Morey has an identity problem, sitting as it does between Gevrey to the north and Chambolle to the south. The former boasts Chambertin as its standard-bearer, the latter Musigny, making it easy to pigeonhole them as masculine and feminine sides of the same coin. Morey doesn't possess a vineyard of such renown, and when the time came to append a grand name it chose Saint-Denis rather than La Roche. Morey was the last village to do this, fully eighty years after Gevrey, in 1927. There's no doubt La Roche is the better vineyard and a great example can stand compare with the neighbours up and down the road. Perhaps being sainted clinched it for Saint-Denis; lip service is still paid to the saints in the Côte d'Or and their names pop up frequently: Clos Saint-Jacques, Clos Saint-Jean, Les Saints-Georges, Saint-Aubin, Saint-Romain, Romanée Saint-Vivant …

The consequence of sitting somewhere between the fortitude of Gevrey and the finesse of Chambolle was that Morey's style was defined by reference to them, but times are changing. Where Clos de la Roche once had to do most of the flag flying for Morey, with some help from Clos Saint-Denis, today the much improved Clos de Tart and Clos des Lambrays are gaining renown in their own right, while also adding lustre to the commune's reputation. From a good domaine, suppleness and balance are Morey's calling cards, leavened by a sterner spicy element to add distinction. A clearer identity has emerged.

Producers

Domaine Arlaud

41 rue d'Epernay, 21220 Morey-Saint-Denis
www.domainearlaud.com; tel: +33 3 80 34 32 65

When tasting at Domaine Arlaud the challenge is to stay focused on the wines and not let your eyes drift upwards to the magnificent roof timbers of the tasting room, crafted by the same firm that restored the roof of Notre Dame in Paris. Cyprien Arlaud, fully in charge here since 2013, having joined the domaine in 1998, explains: 'We have to keep the old crafts, we must preserve the knowledge of old methods, this technique is called *chevrons formant ferme*.'

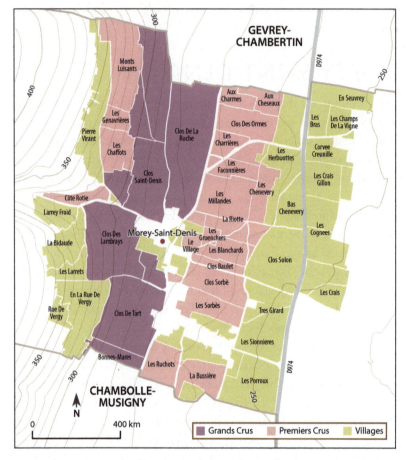

Map 3: Morey-Saint-Denis

Such an observation may have little to do with the wines but it speaks of a mindset and an approach to the broader business of winemaking that is both commendable and rare. Arlaud does not see his winemaking as an activity separate from the whole of life but part of an intricate web of interconnected philosophies, influences and practices. And he is a master of insightful quotes to elaborate his thinking. On walking the vineyards: 'The more time you spend in the vineyards the more you see – I see something different every year, for me it is fascinating.' On the character of Les Ruchots, a Morey-Saint-Denis *premier cru* that lies across the road from Clos de Tart and in which he is the largest owner: 'It is a pure expression of what is a *climat* … It's like a paradise … It's a good place for vines.' On his adherence to organic and biodynamic

THE VILLAGES AND PRODUCERS OF THE CÔTE DE NUITS 93

principles: 'It is very important to have a global approach for the soils, the vines, the people …'.

More prosaically, Domaine Arlaud was founded in 1942, the first vintage to be bottled on site was 1959 and, today, Cyprien represents the third generation in charge of the 15 hectares that include holdings in four *grands crus*: Clos de la Roche, Clos Saint-Denis, Charmes-Chambertin, and Bonnes Mares. He constantly emphasizes the need for full control, especially at harvest time, when his team of 80 pickers is employed directly by him and not through an agency, allowing him to stop and start as necessary, depending on the dictates of the weather.

Once picked by hand, with an initial sorting in the vineyard, the grapes are further sorted on three tables at the winery. The *village* wines are all de-stemmed, while the *premiers crus* and *grands crus* use a minimum of one-third whole bunches. After a cool maceration lasting five days, fermentation tales place in concrete, stainless steel and wood, all wood for the *grands crus*. Barrel ageing lasts for about 16 months with a maximum of 15 per cent new.

The results are compelling, wines of style and substance that engage the senses thanks to an unfolding array of refined flavours. Greater composure has been achieved in recent years, the sometimes brusque flavours of a decade ago now a memory. Another decade at this rate could see them become some of Burgundy's most sought-after reds.

Try this: Chambolle-Musigny

According to Cyprien Arlaud, 'Chambolle is a love story between Pinot Noir and the village.' There are few who would disagree with him. The grapes for this wine come from a cluster of *lieux-dits* located to the north of the commune, on the boundary with Morey-Saint-Denis: Les Bussières, Les Herbues, Les Chardannes and Les Gamaires. The result is an utterly seductive wine, near-impossible to consign to the spittoon, or *crachoir*, thanks to delicate tingly fruit and perky acid that resolve into a harmonious, reverberating finish.

Domaine Dujac

7 rue de la Bussière, 21220 Morey-Saint-Denis

www.dujac.com; tel: + 33 3 80 34 01 00

Websites are tricky things, some seem conceived to dazzle rather than inform, others are so basic they prompt a 'why bother?' response in any who consult them. Not so at Domaine Dujac. This is a proper shop

94   CÔTE D'OR

window first, visually attractive, and then a useful resource; it strikes a good balance between style and substance, just like the wines.

Much has changed at Dujac since the first edition of this book, most visibly the new winery that saw its first vintage in 2022. It is an essay in eco-friendly construction: the 60 centimetre thick walls are built from boxes filled with compressed straw, while the oak beams were sourced from an old building, and Jeremy Seysses reckons they could be five or six hundred years old. Jeremy and his wife, Diana, along with his brother, Alec, have now taken over completely from Jacques and Rosalind Seysses, who helmed the domaine for some 50 years, taking it from new kid on the block in the late 1960s to Burgundian aristocracy by the early years of this century.

Housed within the new building is a set of new stainless steel vats in three sizes: 35, 53 and 73 hectolitres, their USP being that they are double-walled and fully insulated as well as being temperature controlled. They are used for the whole-cluster fermentation that has been the Dujac signature for decades. About 85 per cent is the norm, though Jeremy Seysses explains that the policy is not written in stone: 'I might go one way or the other, depending on the vintage.' In 2022, for instance, the Echézeaux saw 100 per cent whole bunches, while 75 per cent was used for the Charmes-Chambertin.

The grapes are sorted first at the vineyard and again at the winery, where they are also cooled if necessary. During fermentation, in those 'not inexpensive' tanks, pumping over is favoured over punching down and after 10–15 days the wine is pressed in a bladder press, a task made easier by the use of whole bunches. Free-run and press wines are then transferred back to tank to complete fermentation. Jeremy Seysses describes the vinification as, 'a very dynamic process, it's not linear, you plot your course and keep within what you want but our approach is responsive, not dogmatic.' As an aside, he adds: 'Initially the wine might be a little bit funky but not problematically funky.' Élevage follows in standard *pièces*, the percentage of new barrels ranging from about 20 for *village* wines up to about 70 for the *grands crus*. Dujac buy their own staves and then get the barrels made by *tonnellerie* Remond.

Jeremy and Diana's son, Aubert, lent a juvenile hand during harvest 2022 and judging by results, he chose a good year for his maiden vintage. It is a vintage that Jeremy describes as 'comparable but different to 2018; it doesn't have the weight of '18, we took a step back and used a gentler approach.' It was a step in the right direction for, across the

Jeremy Seysses, Domaine Dujac, Morey-Saint-Denis

range, the 2022 wines are defined by ripe fruit and refined abundance in every element. Style matches substance in them all. And, as Jeremy says, 'the wines are sweet without being cloying.'

Try this: Clos de la Roche grand cru

Clos de la Roche is Morey-Saint-Denis' largest *grand cru*, at about 17 hectares, of which Dujac owns nearly two, spread across five parcels and several *lieux-dits*; thus the wine delivers an accurate summation of the vineyard's style and character. The *clos* lies at the northern limit of the commune, on the boundary with Gevrey-Chambertin, and it is not fanciful to suggest that this imbues the wine with more substance than its neighbour, Clos Saint-Denis. But this wine has a character all its own, thanks to a spicy inflection that gives the flavour a focal point. Civil richness and lovely freshness go hand-in-hand with svelte tannins – all the way into an echoing finish.

Domaine Robert Groffier Père et Fils

3 route des Grands Crus, 21220 Morey-Saint-Denis
Tel: +33 3 80 58 54 49

Currently, the name Groffier does not sit high in the Côte d'Or hierarchy but that is about to change for, in the deft hands of Nicolas Groffier,

96 CÔTE D'OR

this domaine is producing wines of rare beauty and style. Though located in Morey, most of the vineyard holdings are in next door Chambolle-Musigny and it is not fanciful to suggest that there is a hand-in-glove fit between Groffier's methods and Chambolle's reputation for vinous elegance. Nicolas' wines burnish that reputation – and then some. Save for the Chambertin-Clos de Bèze (below) the wines all display bountiful red fruits leavened by a will-o'-the-wisp thread of spice and tannin.

In speaking of his methods, Groffier is at pains to emphasize that there is no rule book, particularly when it comes to de-stemming: 'I am totally open to it if it is necessary ... We work for the flavour ... I have no fixed policy or philosophy ... It is a tool and we use the tool.' Thus, for instance, the Chambolle-Musigny 1er cru Les Sentiers is 100 per cent whole bunch, partly because the fruit from the 80-year-old vines is difficult to de-stem. The vineyard lies below Bonnes Mares, on the boundary with Morey-Saint-Denis, and from his 1-hectare parcel Groffier produces a wine that carries some of the structure of the *grand cru*, wrapped in exuberant red fruits. The Bonnes Mares itself is fuller in every dimension, yet still refined, the flavour unfolding slowly as it crosses the palate and only finally revealing itself on the long finish.

He is particularly proud of his Les Amoureuses, to the extent that he serves it after the Bonnes Mares in a tasting and, not only that, he produces two versions of the wine, both called Les Amoureuses, distinguished by a different typeface on the label and, in tiny writing, *La grâce des argiles* on one version and *La délicatesse des sables* on the other. The former is grown on clay soil and accounts for 80 per cent of production, while the latter grows on sand. There is little to distinguish them – both are superb – but for consumers this is a confusing exercise, fascinating rather than convincing, like listening to the violin and piano parts of a Beethoven sonata separately instead of together.

Minor reservations aside, Groffier's sensitive touch in the cellar yields wines of remarkable finesse in the glass. Even the basic Bourgogne Passetoutgrains is excellent, a superb example of this too-seldom-seen burgundy. Make no mistake, Groffier's star is in the ascendant.

Try this: Chambertin-Clos de Bèze grand cru

'We are not in Chambolle now, we take a different approach,' says Nicolas Groffier, when speaking of this wine. The Groffier parcel is a 0.4-hectare narrow strip that runs all the way from the road up to the treeline, climbing about 25 metres in the process. This is made with

THE VILLAGES AND PRODUCERS OF THE CÔTE DE NUITS 97

100 per cent whole bunches and is everything that a Gevrey *grand cru* should be. The nose bounds from the glass, opulent and rich, and then the palate takes over with deep, earthy, though not rustic, flavours of dark fruit, spice and sweet incense. There is substance and style, in that order.

Domaine des Lambrays

31 rue Basse, 21220 Morey-Saint-Denis
www.lambrays.com; tel: +33 3 80 51 84 33

I defy anybody to stand in the new vat room at Lambrays and not be impressed by the magnificence of the space, not just the splendid architecture but also the sense of Cistercian calm. The stillness, the tranquillity, is sensed rather than observed. Domaine des Lambrays has been owned by LVMH since 2014, whose deep pockets created this vat room and, when standing there in silent wonder, it is hard to decry the faceless hand of corporate influence in Burgundy. This is truly a special place.

'You don't make a good wine just because you have a nice winery,' says Jacques Devauges, who made the short journey, literally a couple of hundred paces, from his previous employment at Clos de Tart to take up the reins at Lambrays in 2019. He immediately made the switch to organic farming and noticed a huge change after only one year. Biodynamics followed in 2020: 'Organic and biodynamic viticulture give accuracy and definition in the wines.'

Devauges didn't stop there, setting about a minute analysis of the soils in Clos des Lambrays, the *monopole* that isn't. LVMH owns 99.5 per cent of the *clos*, the remainder belonging to Domaine Taupenot-Merme. 'It is better to celebrate what we have than to cry about the bit we don't have,' says Devauges when asked about this, adding, 'we are ready to buy if they want to sell'. Eleven distinct plots were identified, their demarcation influenced by factors such as geology, soil, exposure and microclimate. Devauges explains, 'there is a higher level of diversity in Lambrays than in other *grands crus*,' thanks to undulations in the ground and a steep slope that rises 60 metres from bottom to top.

He varies the viticulture to suit the different plots: they are farmed, harvested, fermented and aged separately and then 're-unified, not blended' in exact proportion to the quantity each produced. '"Blending" suggests adjustment, I don't vary by one litre what goes into the final wine.'

Before that, the wines are fermented in open-top wooden vats in the new winery, which saw its first vintage in 2022. There are 19 of them,

The new vat room at Domaine des Lambrays

made by François Frères, their distinctive feature being that they are cylindrical, not conical, which allows a stainless steel cover to be moved up and down as necessary within the vat, so that the surface is not exposed. This design was Devauges' own idea and he trialled one vat each from Taransaud and François Frères in 2020 and '21 before settling on Frères. He uses 100 per cent whole bunches with one week's maceration before fermentation.

The result is a wine of real class and not one that 'doesn't sit in the first rank of *grands crus*,' as I wrote previously. There is amplitude and depth, vivid cherry-fruit flavours with a hint of violets, sweet spice, great freshness and echoing length. It sits at the top of a range of wines, greatly expanded in 2022, that includes a pure and elegant Morey-Saint-Denis 1er cru Clos Baulet and an infuriatingly-hard-to-spit Vosne-Romanée 1er cru Les Beaux Monts. Each answers Devauges' stated belief: 'Terroir must be upfront, technique behind. Musicians transmit emotions, if you feel their technique it breaks the emotion. If wines are too technical it is a disaster.'

Try this: Nuits-Saint-Georges 1er cru La Richemone
Richmone is a seldom-seen Nuits *premier cru* of two hectares that lies on the mid-slope to the north of the appellation, a couple of hundred

metres from Vosne-Romanée. Any suggestion that this proximity influences its character is met with a firm, 'No, it has its own style,' from Jacques Devauges. The wine might best be described as a traditional Nuits that has had a makeover; the best of the old style has been retained, by way of a grippy tannic foundation, but any rough and rustic elements have been refined and polished. There is still a firm structure but it simply provides the support for the elegant fruit and juicy acidity that overlay it. This is a superior Nuits.

Domaine Georges Lignier et Fils

41 grande rue, 21220 Morey-Saint-Denis
Tel: +33 3 80 34 32 55

The original Georges Lignier was followed in turn by his son and then grandson, also Georges, from whom his nephew, Benoît Stehly, took over in 2012, having worked alongside his uncle since 2002.

Stehly talks animatedly, whether it is about the multi-tasking challenges of being a vigneron today, '30 years ago you could just be a winemaker, now you have administration, marketing, advertising …' or the style he is aiming for: 'I want Pinot Noir that is able to age thanks to the freshness. I don't want heavy wine … Sometimes a big wine is good for one glass and then boring. I don't want wine to be boring. It must be kept fresh … I don't want to make wine the same as everybody else, I want to make Lignier.'

If the above can be called a mission statement then he is true to it, for the house style is racy and distinctive, individual, with lean and focused flavours and no rich textures. 'Freshness' is Stehly's one-word mantra, used repeatedly, no matter what wine he is discussing. But when freshness becomes a euphemism for austerity it possibly needs to be reined in a little; there is a fine dividing line between the two and some of the Georges Lignier wines threaten to cross it.

The Morey-Saint-Denis sports a dry bite of tannin and fresh bitter cherries; the Gevrey-Chambertin is cut from a similar cloth, with a rustic rasp; the Chambolle-Musigny is equally firm and tannic. In all three cases, five years' ageing will polish the edges into harmony, so these can only be provisional observations, particularly in a textural sense. In some respects the trio should be celebrated, both for their individuality and for the fact that, today, everybody is seeking to deliver immediate charm. It is nice to encounter a vigneron with an eye on the future too.

100 CÔTE D'OR

The *grands crus* have the fruit weight to support the house style, adding welcome breadth to the *village* wines' linear profile. There is a full hectare of Clos de la Roche, scattered across five plots, and nearly one-third of Bonnes Mares, in a narrow strip that runs almost the full width of the vineyard. The former's ample fruit balances the firm texture harmoniously, while the latter delivers a full-bodied wave of flavour before resolving into a delicate finish.

As with many Côte d'Or winemakers, Stehly has no fixed policy with regard to de-stemming, usually using between 30 and 40 per cent whole bunches, depending on the vintage. 'We used 40 per cent in 2014 and it was too much,' he notes. 'We could have used less to get more fruit flavours. It is a tool we have to be careful with.'

Try this: Clos Saint-Denis grand cru

Clos Saint-Denis is named after the first bishop of Pars, who met a nasty end around AD 250 when he was beheaded. It should not be confused with the tiny *lieu-dit* of the same name at the southern end of Echézeaux. Lignier is the biggest owner in the *clos* with 1.49 hectares in two parcels, a whisker more than Domaine Dujac with 1.47. The Lignier plots lie at the northern and southern end of the vineyard and in total they equate to about one-quarter of the clos. The wine cleaves to the house style, with leaner elements to the fore. Crisp acidity and firm tannins lead the fruit into a long finish. By comparison with Lignier's Clos de la Roche this is sterner and less opulent.

Domaine Hubert Lignier

13 route Nationale 74, 21220 Morey-Saint-Denis
www.hubert-lignier.com; tel: +33 3 80 51 87 40

'It is important to have experience of different estates,' explains Laurent Lignier when speaking of his son, Sébastien, having worked in Côte Rôtie, Oregon and Valle d'Aosta as well as studying at the 'Viti' – the Lycée Viticole de Beaune. Father and son represent the fifth and sixth generations at this quality-conscious domaine, which is not noted for shouting its wares. The wines do the talking.

In that vein Laurent Lignier speaks with conviction bereft of dogma when discussing his winemaking and the family domaine, which encompasses 11 hectares of vineyard. These lie principally in the Côte de Nuits with some further south in the Côte de Beaune, including parcels in Pommard, Volnay and Monthelie bought with friends in 2020. The domaine has been organic since 2008 and certified since 2019, with some biodynamic practices also implemented.

He doesn't have a rigid de-stemming policy and after sorting on a vibrating table he favours a six- or seven-day cool maceration before fermentation with natural yeasts. *Remontage* is favoured over *pigeage*, especially early in the fermentation. Vinification lasts for about 18 days, with the press wine added back for a week in tank, 'to assemble,' before élevage of about 20 months in barrel. These are sourced from seven or eight coopers, mostly François Frères and also Taransaud and Remond because, 'I like the diversity,' says Lignier. The wines are racked by gravity and bottled without fining or filtration.

A signature of the Lignier wines is the high proportion of old vines, in such as the Gevrey-Chambertin Les Seuvrées, which is made from Laurent's grandfather's plantings (1947) and his father's (1968). It is a suave, juicy mouthful, though the Morey-Saint-Denis 'Trilogie' surpasses it thanks to greater underlying power and grip. As the name suggests, it is made from three *lieux-dits*: Les Chenevery (planted 1936–44), Clos Solon (1966) and Les Porroux (1960–80).

Lignier's holding of almost 1 hectare in Clos de la Roche, divided two-thirds, one-third between Monts Luisants and Les Fremières, allows him to make two cuvées, which he has done since 2017. His grandparents bought the holding in 1944 and the vines date from 1955. The 'standard' bottling is an elegant beauty and the Cuvée Hommage, made with small berries affected by *millerandage,* is a worthy tribute to his grandfather. Three barrels were produced in 2022. It is a wine of initial quiet restraint that opens and unfolds gloriously: there's spice, there's savour, there's fruit, all interlaced seamlessly and continuing into a reverberating finish. As a standard-bearer for the house style it could not be bettered.

Try this: Chambolle-Musigny 1er cru Les Baudes
The name may derive from *bode*, meaning a dwelling place (abode) and the vineyard lies north of the village, almost on the boundary with Morey-Saint-Denis. Lignier's 0.2-hectare holding is easily found at a fork in the road marked by a cross and also because the vines, dating from 1957, are planted on a north–south orientation. This has protected them in recent hot vintages, meaning that the roughly three barrels he makes annually contain a wine of delicious, immediate appeal. The acidity is fresh, the tannins are soft and the finish is long.

102 CÔTE D'OR

Domaine du Clos de Tart

7 Route des Grands Crus, 21220 Morey-Saint-Denis
www.clos-de-tart.com; tel: + 33 3 80 34 30 91

Some Côte d'Or domaines are fiendishly difficult to find, but not Clos de Tart, for the name is emblazoned, large and eye-catching, on the street wall of the property. Despite that flourish everything else here looks inward, towards the winery and vineyard, which dates from 1141. It was bought that year by the nuns of Notre Dame de Tart, a Cistercian order for women, who wore the same habit as the monks along with a black veil, and it remained in their possession until the Revolution. The nuns' centuries-long custodianship ended abruptly when they were expelled and the *clos* was sold as a *bien national* in 1791 to the Marey, subsequently Marey-Monge, family. It flourished during the nineteenth century and was ranked as one of the Côte d'Or's best, along with the likes of Chambertin and Musigny.

Mommesin bought Clos de Tart in 1932 and over the following decades gradual decline set in; being shackled to the humdrum négociant side of Mommesin's business did the *clos* no favours. Boisset bought the négociant in 1996 and revitalization of the *clos* itself followed almost immediately under the careful stewardship of Sylvain Pitiot, cartographer of the Côte d'Or. In 2018 Clos de Tart changed hands again when it was acquired by Artémis Domaines (which also owns Bouchard Père & Fils, p. 168, Château Latour, Château Grillet and others). Thus it has had only four owners in its almost 900-year history.

Alessandro Noli is now the man at the helm, arriving in March 2019, having worked previously at Latour and Grillet. He immediately took charge of a massive programme of renovation that lasted from January 2019 to May 2022: 'We did the vat room, the cellars, the offices – everything!' The vineyard itself was minutely analysed and mapped, identifying 16 different plots in the process: 'We brought a real focus on the terroir to understand it,' he recalls.

Sixteen new wooden vats were bought, their capacities corresponding to the size of each plot: one of 40 hectolitres, three of 35, diminishing down to four of 20 hectolitres. 'Each cuvée gives a different expression, so we take a different approach for extraction with each,' says Noli, 'we don't need to extract too much, it will come naturally.' After fermentation, a new vertical press is used – 'it gives more elegance' – and each cuvée is transferred to barrel and kept separate for seven to eight months, then transferred to stainless steel tank for blending of the final cuvée, before going back to barrel for the remainder of the élevage. 'They have to grow up together,' says Noli, 'the magic is when you blend'. Less new oak is used than previously, down from 100 per cent to about 60: 'In some of the old vintages the structure of the tannins was a little rustic,' he notes.

Noli, who uses about 50 per cent whole clusters for Clos de Tart, is currently engaged in what can only be described as a huge research project, in conjunction with Bordeaux University, on the use of stems. Scheduled to run for four years, it involves massive data collection and would be worthy of a PhD thesis. Noli takes stems seriously: 'Whole bunch is to support, not lead, the terroir. Too much will cover the terroir.' It is safe to say that Clos de Tart is well on its way back to regaining its nineteenth-century standing as a *grand cru* of first rank.

Try this: Clos de Tart grand cru

The *clos* is the largest of the Côte d'Or's five *grand cru monopoles* and the only one outside Vosne-Romanée. In addition, its 7.5 hectares form a rough square and the vine rows run north–south rather than the more usual east–west. It is a place apart. Most pertinently, the days of ho-hum quality are now historical, banished by the ever more meticulous approach of recent years. A spicy perfume presages a palate of restrained power and richness, suave tannins have replaced the rustic ones of old, and there is great conviction on the impressively long finish.

Domaine Cécile Tremblay

8 rue de Très Girard, 21220 Morey-Saint-Denis
Tel: +33 3 45 83 60 08

New Year's Day – *Jour du l'An* – 2022 was a red-letter day for Cécile Tremblay, when her domaine doubled in size thanks to the expiring of long leases on family vineyards, which then came under her control. About 4 hectares were added to her holdings, including a quarter in Clos de Vougeot, one of *village* Morey-Saint-Denis and a relatively whopping, 1.8-hectare parcel of the prized Vosne-Romanée *premier cru* Les Beaux Monts. Father Christmas had arrived a little late, but with what bounty.

An extension to the cellar to handle the extra fruit has recently been completed, including a cool room for the grapes: 'That was the missing piece,' says Elodie Bonet, Tremblay's assistant. Prior to that, refrigerated

104 CÔTE D'OR

trucks had been hired for this purpose, a practice more common than might be supposed.

This is a young domaine – Tremblay only started in 2003, yet she quickly earned a name for herself as a winemaker of rare sensitivity and skill. Finesse and intensity, allied to a haunting elegance, made the wines utterly seductive; they were to be revelled in first and analysed later. More recently the style has matured, adding gravitas to grace. The elusive, tantalizing flavours that characterized the wines have yielded to a more easily discernible character, centred on a darker fruit profile. There is more amplitude in the texture.

A good example is the Bourgogne Côte d'Or, replete with plump dark fruit and liquorice, sumptuous and enduring on the finish. Further up the scale the Chambolle-Musigny 1er cru Les Feusselottes is full-bodied with a dense depth of flavour; while the Echézeaux *grand cru*, from a parcel in the du Dessus *lieu-dit*, is a hearty wine, concentrated and full.

All the wines are treated to kid-glove oak treatment in Chassin barrels, made to order for each wine, the oak being sourced from several forests, named on the barrels: Perseigne and Jupille, Compiègne and Prieuré, for example. A trio of 400-litre ceramic barrels are also used to emphasize the fruit flavours in the Bourgogne Côte d'Or, the Morey-Saint-Denis and the Beaux Monts. There is no set de-stemming policy, it varies according to vintage and vineyard, and fermentation is in wood and stainless steel. Even when whole bunches are used the main stem is removed because it adds coarse tannin when not optimally mature. Pumping over is favoured over punching down; a post-fermentation maceration follows and after pressing the wines spend 15–18 months in barrel before bottling and closing with natural cork. 'We try to touch the wine less,' says Bonet.

Try this: Vosne-Romanée

Grapes for this wine come from two *village* parcels: Aux Communes and Les Jacquines. The former abuts the D974 and is divided roughly one-third, two-thirds by a tree-lined avenue leading to the heart of the village; while the latter, the name of which references a previous owner, M Jacquinot de Richemont, a parliamentary representative from Dijon, lies close to Nuits-Saint-Georges. Harmony and balance define this wine; fruit, acidity and tannin are woven into a delicious whole to yield a vital surge of energy on the palate.

CHAMBOLLE-MUSIGNY

There are few names more evocative of Burgundian grace and beauty than Chambolle-Musigny; even in saying it there's a ring of pleasure, a ring answered in abundance by the wines themselves. While it doesn't surpass Vosne or Gevrey in outright status, Chambolle is held in special affection by all burgundy drinkers. Perhaps it starts with that soft and flowing name?

Chambolle is a discreet village, a neat collection of houses tucked into the hillside like a baby in a cot, much like Volnay in the Côte de Beaune. It is not just their wines that are often likened to one another, the villages, too, share much in common in terms of situation and scale. Both sit a little further up the hillside than their immediate neighbours, set back from the main road, from where each is easily identified, Volnay by the façade of Pousse d'Or and Chambolle by the distinctive tower of the village church. The wines call for attention like no others, challenging the professional taster to make an arms-length, objective assessment of their qualities. They must be engaged with, not observed. The gravitas of Gevrey is not found in Chambolle, nor the majesty of Vosne; the wines are less richly endowed but lack nothing in attraction for that. If beauty was the only criterion Chambolle would rank above all others.

It is in Chambolle that Pinot Noir's signature qualities of finesse and delicacy, with no lack of intensity, shine brightest. Here is where the match of grape and ground approaches perfection, where Pinot's qualities are supported and amplified, not dominated, by the terroir and where the vine in turn transmits the terroir message with greatest facility. The duet between grape and ground is all about harmony and synergy, it's a hand-in-glove fit. There's strength without overt power, vigour without excess weight. The village lies between its two celebrated *grands crus*, Bonnes Mares and Musigny, which stand sentinel to north and south, each on the boundary with the neighbouring commune. There is also a clutch of two dozen *premiers crus*, led by the famed Les Amoureuses, which lies on rugged terrain downslope of Musigny.

Producers

Maison François Millet & Fils

15 rue du Carré, 21220 Chambolle-Musigny

Email: mfmillet@orange.fr

François Millet retired as winemaker at Comte Georges de Vogüé after his final vintage in 2020, a role he had filled since 1986. During his long tenure he came to be known for his lyrical, sometimes mystical, descriptions of the wines, a talent that earned him the sobriquet 'The Poet of Burgundy'.

Millet now devotes himself to making wine in his own name along with his sons, Julien and Adrien, in the cellars beneath his home, within shouting distance of his former employers. And he hasn't lost his touch, either with words or with wines: 'If you push you disfigure the wine, you must be diplomatic as if you are dealing with a thin-skinned person. Guide the wine to respect the potential, don't try to get too much.' The language might be obscure at times but it is far more engaging and memorable than a litany of techno-babble, while the wines speak for themselves with pure flavours uncluttered by intrusive winemaking.

If ever a winemaking operation could be described as 'boutique' this is it. Annual production has averaged about ten thousand bottles since the first vintage in 2017, made before Millet retired. He had hoped to make some wines from 2016 but such were the depredations from spring frost that year that no grapes were available for him to buy. Acknowledging the tiny scale of operations, Millet avers, 'We don't do pump over, we do bucket over.' Yet all the pieces are in place, from sorting table to bottling line, to Stockinger barrels – 'you don't see the oak', he notes. The Millets are not cutting corners.

Village decorations in Chambolle-Musigny for the Saint Vincent Tournante 2024

So far, so good. The Achilles heel in such a set-up, however, is that you are at the mercy of the growers you buy from and, while Millet has good relationships with the half-dozen suppliers, he has no control over the available quantity, as in 2016, and also 2021 when he had to hunt about for other grape sources.

Those grapes are always 100 per cent de-stemmed, a policy that Millet describes as a pragmatic choice rather than a dogmatic principle, elaborating: 'Whole bunch puts a veil between the wine and the land, I prefer to de-stem. The connection with the land is decreased by stems. I want the true identity of each lot, the precise definition of place.' All his 15 wines chime with that approach and if one can be singled out as representative of its origin it is the Bourgogne Rouge. It comes from a plot at the bottom of the slope opposite Chambolle-Musigny and is best described as an outline sketch for a fine Chambolle, thanks to its sweet perfume and ripe fruit.

Millet continues: 'You must be careful, you want to enhance the place. Be cautious, just make an infusion. It is good to converse with the land, some is introverted, some extroverted, like people. We don't want to impose anything on the land.' François Millet has lost none of his lyricism.

Try this: Volnay 1er cru Les Angles

Volnay and Chambolle-Musigny are often considered as cousins from across the divide that separates the Côte de Beaune from the Côte de Nuits, each extolled for wines of grace and harmony, never harsh or rugged. It seems appropriate, therefore, that Millet, who spent his working life in Chambolle, should now be turning out Volnays that cleave to the classic template. There is a delicious Les Grands Poisots and this splendid *premier cru*, which is fine and precision-flavoured with 'a certain spiciness,' according to François, 'and velvet tannins,' adds his son Julien.

Domaine Jacques-Frédéric Mugnier

Château de Chambolle-Musigny, 2 rue de Vergy, 21220 Chambolle-Musigny
www.mugnier.fr; tel: +33 3 80 62 85 39

Musigny: half a dozen landholders in this renowned *grand cru* make do with holdings of less than a quarter hectare but not Frédéric Mugnier, the second-largest owner (after de Vogüé) whose holding amounts to 1.14 hectares. It comprises a broad band in the middle of the vineyard, a smaller block at the northern end and a tiny sliver close by that is barely visible on a large-scale map. Mugnier crafts a magnificent wine from

108 CÔTE D'OR

those gentle slopes, one that he feels so strongly about that he no longer releases it with his other wines but holds it back for five or more years to encourage people to drink it when properly mature.

Mugnier explains: 'It is important to do it and it is absolutely necessary. Otherwise, people have no idea of the real thing, of what makes a *grand cru* so different, its deep potential,' adding that customers didn't dare complain when apprised of the new policy. He is in the fortunate position of being able to do this with his most prestigious wine thanks to having relatively abundant quantities of his Clos de la Maréchale (below) to keep impatient customers satisfied in the short term.

Though the domaine can be dated back to 1863 Mugnier was not born with burgundy running in his veins. He boasts a more cosmopolitan background than many Côte d'Or proprietors and is far removed from the grizzled peasant stereotype of yore. He grew up in Paris, only spending holidays in the family's Château de Chambolle-Musigny, an impressive property at the top of the village. At that time the vineyards were all rented out and it was only after his father took some of them back in 1977 that Frédéric evinced some interest in winemaking as a career.

It is forty years since Mugnier took over from his father and he is now one of the most highly regarded winemakers in Chambolle-Musigny and, by extension, the Côte d'Or. Despite the domaine's diminutive size he acknowledges how lucky he was that it included some magnificent vineyards, with holdings in Bonnes Mares and Les Amoureuses as well as the Musigny. 'Harmony' is a word he uses frequently when discussing his wines and if the house style could be described in one word that would be it. He thinks deeply about his winemaking and is no fan of the dogma or restrictions inherent in a rigid application of organic or biodynamic principles. Indeed, any who are keen to expand their understanding of the latter should read his *A Closer Look at Biodynamics*, which is posted on the website above. But do not ignore the wines, they speak most clearly for him.

Try this: Nuits-Saint-Georges 1er cru Clos de la Maréchale

'I like this vineyard, it gives me some rest from the others,' says Mugnier, and thanks to his sensitive winemaking hand this *monopole*, the largest in the côte at 9.5 hectares, produces a wine best described as a Nuits in Chambolle clothing. Refinement and elegance go hand in hand with dark fruit and some spice, with no sense of conflict between these contrasting elements. It is vibrant and vividly flavoured without being

THE VILLAGES AND PRODUCERS OF THE CÔTE DE NUITS 109

overbearing. There is also a small amount of white made which, despite the plump fruit profile, manages to avoid the oiliness sometimes found in Côte de Nuits whites, and is all the better for that.

Domaine Georges Roumier

4 rue de Vergy, 21220 Chambolle-Musigny
www.roumier.com; tel: +33 3 80 62 86 37

Christophe Roumier represents the third generation of his family at this domaine, following his father Jean-Marie and grandfather Georges, who founded the domaine in 1924. He started working alongside his father in 1981 and took over completely in 1984. He has now been joined by two nephews, Alexis and Clément, and observes, 'It is nice to see that there is a future for them in winemaking. It was less attractive for my generation, you can make money now.' He is beginning to stand back from his hands-on role of 40-plus years: 'I am now advising the younger people, they are taking on the work.' And they are lucky to have such a mentor to call on, for Roumier is one of the most highly regarded of all Côte d'Or vignerons.

He has been working organically since 1993 but is not biodynamic, saying, 'I don't think it makes a difference. If I had a full understanding I might change. It is a concept, more philosophy than anything else.' His tone is observational, this is not a trenchant dismissal and that is Roumier's secret, dogma plays no role in his approach. When asked about whole bunches or de-stemming he simply replies that it depends on the vintage, though as the years have passed he has tended to use more. He likes whole clusters in riper vintages, such as 2022: 50 per cent in Chambolle-Musigny *village*, 60 per cent in Morey-Saint-Denis 1er cru Clos de la Bussière, and 80 per cent in Chambolle-Musigny 1er cru Les Cras and Bonnes Mares *grand cru*.

Fermentation is in stainless steel or concrete: 'I like the two, the size of the vat makes the choice.' The vats are initially cooled to 15°C and held at that for about a week, which allows him to complete his harvest before he returns to the winery, pumping over the vats, 'to encourage fermentation'. He has almost stopped punching down because, 'tannin extraction is easier than in the 1980s,' and he uses less new oak than previously, stating, 'I never loved oak, it was helpful when tannins were less ripe but it is not needed so much now.' Almost all his barrels come from François Frères in Saint-Romain and he uses modest percentages of new barrels, up to 20 per cent in the Bonnes Mares. The wines are racked out of barrel

110 CÔTE D'OR

at the end of the year following harvest, assembled in stainless steel for two to three months and bottled in March, some 18 months after harvest. Roumier concludes: 'It is an ordinary process, nothing special.'

Ordinary it may be but the results are far from ordinary, as legions of burgundy lovers across the globe will attest. His Musigny, from a 0.1-hectare plot in the north-western corner of the vineyard, attracts frenzied interest on the secondary market, where it changes hands for crucifying prices. To experience the Roumier magic, however, it is not necessary to mortgage the house: his Chambolle-Musigny 1er cru Les Cras is a magical wine that thrills the senses, from the sweet perfume with a mild exotic note, through the resplendent palate of gorgeous fruit and gentle spice, to the slight saline lift on the finish. Roumier describes it modestly: 'There is expressive purity and transparency, it is a good approach to Bonnes Mares, as it has the same profile.'

Try this: Bonnes Mares grand cru

Roumier's Musigny enjoys such fame that it tends to overshadow this sibling *grand cru* from the other, northern, end of the Chambolle-Musigny commune. At nearly two hectares Roumier's is the second-largest Bonnes Mares holding, after de Vogüé, and is spread across a quintet of plots that combine the two geologies of the vineyard: the white soil of the upper part and the brown of the lower. Save for a small quantity of wine made from each for the purposes of comparison, the domaine produces one wine blended from all the components. It is a wine that challenges the tasting lexicon: there is spice and savour, a gorgeous rasp of fruit, there's depth, complexity and length. Words fail.

Domaine Comte Georges de Vogüé

7 rue Sainte-Barbe, 21220 Chambolle-Musigny

Tel: +33 3 80 62 86 25

The Côte d'Or is host to many domaines with a multi-generational history but few can trace their origins back to 1450, as Domaine Comte Georges de Vogüé can. There is a patrician feel here and the premises, though large by Côte d'Or standards, are discreet and access is through an arched passageway that opens into a trim courtyard. There's a reserved air about the place, which mimics the stereotypical Bordeaux château, where the owner is seldom seen and visitors are hosted by the management or winemaker, in this case Jean Lupatelli, who succeeded the long-serving François Millet in 2021.

Jean Lupatelli, winemaker at Domaine Comte Georges de Vogüé, Chambolle-Musigny

That vintage delivered a baptism of fire for the newly appointed winemaker, though ice might be a more appropriate metaphor for such a troubled year. Right across the côte April frost dealt a devastating blow to the delicate new buds, just as it had done in 2016. Yet nothing could dampen Lupatelli's enthusiasm for his new job, where he is making, 'no revolutionary interventions, just soft changes.' One of these directly addresses an observation I made in the first edition of this book, though I make no claim for having inspired him: 'There's a reticence about the wines that makes them more challenging than others. Time must be taken with them.'

Lupatelli is conscious that in the best vintages the de Vogüé wines can 'close up between birth and 20 years,' and wants to 'open the window of drinking, the window of emotion.' He now uses 50 per cent whole bunches, where previously there were none, and is bringing 'a more detailed approach' to the winemaking of the flagship Musigny. Specifically, this has involved identifying seven different terroirs within the domaine's large holding (7.12 hectares out of 10.85) and vinifying them as separate cuvées. He concedes that the vineyard is, 'pretty homogeneous,' but still feels that because of subtle differences in aspect, elevation and soil it is a worthwhile exercise, for there are marked

112 CÔTE D'OR

differences in flavour when tasting from barrel. Might he be tempted to bottle them separately? 'I trust too much in the *Climats de Bourgogne* to do that,' he replies.

Lupatelli has also addressed himself to the issue of vine age across the domaine, concluding 'it was time to rebalance' the overall mix of young and old. Some small plots were replanted but more significantly a quarter-hectare of the domaine's half-hectare holding in Les Amoureuses was uprooted – to lie fallow for six years before replanting.

Lovers of Les Amoureuses – a glorious medley of sweet and savoury components – will find it difficult to source for several years because of this. They may have to content themselves either by trading up to the Chambolle-Musigny *premier cru*, which is 'Musigny in mufti', for it is declassified young vine Musigny and could be labelled as *grand cru*; or by trading down to the Chambolle-Musigny *village*, though 'down' in this case indicates a drop in official status rather than quality. The latter comes principally from 1.8 hectares in the *lieu-dit* Les Porlottes, enriched by a small quantity from two *premiers crus*, Les Baudes and Les Fuées, both of which are neighbours of Bonnes Mares. Porlottes faces east and is the last of the de Vogüé vineyards to be harvested. It yields a remarkably expressive wine, simultaneously rich and lean with no sense of contradiction on the palate.

'Why waste such incredible material?' asks Lupatelli rhetorically when explaining that he has also recommenced the production of Marc and Fine de Bourgogne, which stopped in 2013. From barrel, both are predictably incandescent. Stick to the Musigny.

Try this: Musigny Blanc grand cru

This is a wine almost of fable and became even more so in the period between 1993 and 2015 when it was not produced, after the vineyard was replanted and the produce of the young vines was declassified into a Bourgogne Blanc of knee-weakening price. It is an intriguing wine, far removed from the white glories of the Côte de Beaune, marked by a dry rasp at its core that is almost tannic, a slight arrest in the flavour flow that adds more interest than charm. Save for a citrus cut, non-fruit flavours abound in the shape of mild spice and a saline snap on the finish. The 2017 and 2019, drunk side-by-side in 2023, were noted for lovely intensity, though if served blind their identity would probably have remained beyond reach.

THE VILLAGES AND PRODUCERS OF THE CÔTE DE NUITS   113

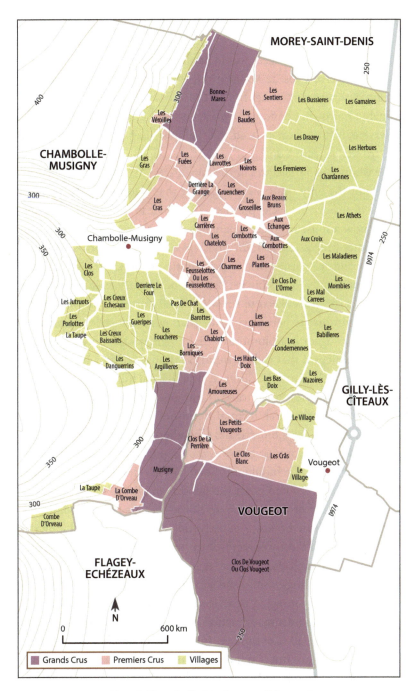

Map 4: Chambolle-Musigny and Vougeot

VOUGEOT

As a wine commune Vougeot would hardly exist were it not for its eponymous *grand cru*, Clos de Vougeot. Here the oft-cited pyramidal progression of vineyard ranking, from a broad base of *régionale* and *village*, through a narrowing band of *premiers crus* and finally to a pinpoint of *grands crus*, is turned on its head. A little more than 3 hectares of *village* vineyard sits beneath a dozen of *premier cru*, which in turn is topped off by the whopping 51 hectares of the famed *clos*.

It is not the only thing about Vougeot that is lopsided; the name is steeped in history yet lacks prestige. The Château du Clos de Vougeot itself is now a museum piece where once it was a working winery, and the summer procession of tour buses creeping up the narrow road towards it confirms this as the historic centre of the Côte d'Or, a place on every burgundy lover's bucket list. The buses carry legions of visitors from across the globe, keen to see where it all started, if not to buy the wine. The visitors are here to view the past, not experience the present.

Clos de Vougeot is an impressive sight, perhaps the only expanse of Côte d'Or vineyard that calls to mind Bordeaux's Médoc, barely undulating as it stretches into the distance. The château hunkers low amongst its surrounding vines, unadorned to the point of gauntness, with no fancy turrets or other architectural flourishes. Built by the Cistercians and now home to the Confrérie des Chevaliers du Tastevin, there could hardly be a greater contrast between the original and the current occupants, one austere, one bibulous, united by their devotion to the grape.

If only the wines could boast the smooth amplitude of the vineyard. Instead, too many of them come up short, perhaps not disappointing, but too often a little hollow, not delivering the satisfying and long flavours that one can rightly expect from a *grand cru*. If ever the producer's name counts above all else it is here. A good Clos de Vougeot is a worthy ambassador for its origin, solid and substantial and rewarding of repeated sipping. The encouraging news is that a greater number now march to that beat, the humdrum reputation increasingly out of step with reality in the glass. It will take time yet for Vougeot's standing to match its historical importance but it is moving in the right direction. The village itself is linear in layout and runs parallel to the D974 road, which bypasses it. The river Vouge, which is more of a substantial stream, rises in the Petits Vougeot vineyard and flows through the village, lending its name on the way.

Producers

Domaine Bertagna

16 rue du Vieux Château, 21640 Vougeot

www.domainebertagna.com; tel: +33 3 80 62 86 04

The Bertagna name dates from the 1950s, when Claude Bertagna, who previously owned vineyards in Algeria, founded the domaine. At one time it encompassed an enviable roster of *grands crus* but most of those had been sold off prior to his selling up completely to the Reh family from Germany in 1982. Eva Reh has been in charge here since 1988 and despite the *grands crus* sell-off the domaine still claims a quintet of holdings in Chambertin, Clos de Vougeot, Corton-Charlemagne, Corton Les Grandes Lolières and Clos Saint-Denis. Eight *premiers crus*, including almost a hectare of Vosne-Romanée Les Beaux Monts, make for an impressive roster of holdings.

All the vineyards are cultivated organically, though not certified, and a recent innovation has been the planting of Asian radishes between the rows after harvest, then removing them before flowering: 'We don't want a vegetable patch,' says Reh. What they, her vineyard manager Arnaud Lecoeur and winemaker Franck Ambeis, do want is to reinvigorate soil that has become too compacted and lacking in microbial life.

After hand harvesting the grapes are sorted at the winery, with a varying proportion of whole bunches used, and *remontage* is generally favoured over *pigeage* for extraction of colour and tannins. Élevage lasts 12 to 18 months in standard Burgundy *pièces*, the percentage of new barrels varying between 30 to 100 per cent depending on the appellation and the vintage.

Ambeis' first vintage in charge at Bertagna was 2023, having worked at the domaine for the previous seven years. Reh declares herself, 'super happy that I offered him the job,' and the results, when tasted from barrel in February 2024, spoke for themselves. The season was a tricky one with an abundant crop that called for a strict green harvest, while the worry at the end of August was a lack of ripeness, the grapes only showing a potential 9 per cent alcohol. Yet three days of September sun saw that jump by two percentage points.

The *monopole* Vougeot 1er cru Clos de la Perrière 2023 was already showing promise in barrel by way of a soft and supple texture and good length. The 2.2-hectare vineyard faces the Château du Clos de Vougeot

Domaine Bertagna's tiny holding in Les Amoureuses: from such unprepossessing origins comes wine of great beauty

and is easily spotted thanks to a gaudy red-on-white sign that forms an arch over the entrance gate. The wine repays a decade in the cellar when the primary flavours yield to more complex notes of *sous bois* and the textures fatten to plumpness. It is a below-the-radar *premier cru* that is well worth seeking out.

Try this: Chambolle-Musigny 1er cru Les Amoureuses
Planting one's feet on the ground anywhere in the Côte d'Or is always a useful exercise, though never more so than in Les Amoureuses, the most celebrated *premier cru* in Burgundy, along with Clos Saint-Jacques. Walking out from the Bertagna winery you follow the route of the Vouge stream, climbing up towards Musigny. The terrain is rough and rugged, almost scrubland. How a wine as beautiful as Amoureuses can come from this virtual moonscape is a source of constant wonder. The Bertagna parcel is less than one-tenth of a hectare on a rocky terrace that produced 200 bottles in 2020. The wine is remarkable: a *sauvage* nose is echoed on the palate, where a herbaceous note leavens the lush fruit, all supported by a discreet tannic underlay.

Maison Mark Haisma

4 rue des Cerisiers, 21640 Gilly-lès-Cîteaux
www.markhaisma.co.uk; email@markhaisma.com

Former United States president Calvin Coolidge is famously quoted as saying that nothing beats persistence and Mark Haisma is the living embodiment of that wise counsel. Persistence has paid off handsomely for Haisma but be in no doubt that this is persistence of nuclear intensity; only unswerving commitment has got him to where he is today. And there is no slacking or easing into a soft life. He lives 'above the shop' in a small apartment he has created for himself: 'I am a simple man, no fast cars, or Burgundy roof tiles.'

The shop in this case is the winery, *sans* roof tiles, that was built from scratch in time to receive its first harvest in 2017 and which has since had to be extended to cater for the expansion in Haisma's production. Today, it has a capacity of 250 barrels, 80 per cent of which is utilized. Haisma wonders aloud what he will do should he reach the 250 limit: 'Then what? Where do we go from there? Do I even want to?'

Putting the future aside, today he speaks as if on a mission: 'I've spent a lot to have the right kit. We're doing the work that needs to be done ... Humility is everything, your "self" is not important. You can't find the beauty if you are not humble. The vineyard is speaking, not you.' The challenge is to keep up as he elaborates on his thinking and practices, at times speaking over his shoulder or from the top of a ladder as he draws samples from the barrels, stacked five-high in places. Interspersed amongst the torrent of quotes are some choice, unprintable ones that suggest the wines might boast a brash, hollering flavour profile. Nothing could be further from the truth: Haisma is currently turning out an assured range of wines that speak of a deft hand and a sensitive touch.

The range comprises 18 wines, the grapes for which come from his own 4-hectare holding plus bought-in grapes from another 6. New since the first edition of this book is a pair from Pernand-Vergelesses: a white *premier cru* En Caradeux and a red *village* Les Pins. Leaving aside the colour difference there could hardly be a greater contrast between the two: the vineyards face in opposite directions, the Caradeux looks east, directly across to the Hill of Corton, while the Pins is more isolated, at the top of a steep slope and facing west. The Caradeux leans towards the crisp, citrus, mineral flavour register, with a waft of tropical

118   CÔTE D'OR

fruit to keep it balanced; while the Pins is vigorous and cherry-fruited. Attaining ripeness is crucial: 'If you pick early you'll make something miserable,' says Haisma.

There is nothing miserable about any of Haisma's wines, mainly because of his attention to detail and hands-on involvement at every stage, especially during harvest, when he is fastidious about sorting out sub-standard berries that might otherwise corrupt the good fruit in the 15-kilogram boxes he uses: 'I pick with the team, I don't sit in the winery.'

Try this: Morey-Saint-Denis 1er cru Les Chaffots

The Chaffots *lieu-dit* covers four hectares and stretches across three *climats*, taking in a chunk of Clos Saint-Denis and a sliver of Clos de la Roche, with the upslope section of 2.6 hectares designated *premier cru*. Haisma describes Chaffots as 'a mighty vineyard,' and mentions that the 2023 vintage was 'not happy when first in barrel,' but then settled nicely. The first impression is of a gorgeous, juicy attack before the palate notes a firm tannic framework supporting an impressive surge of fruit.

Domaine Alain Hudelot-Noëllat

5 Ancienne RN74, 21220 Chambolle-Musigny
www.domaine-hudelot-noellat.com; tel: +33 3 80 62 85 17

Though the address here says 'Chambolle-Musigny' you will search in vain for this domaine in that village. The easiest thing is to say that it lies in the commune of Chambolle but the village of Vougeot, though well removed from the centre. The reason for the confusion is that the boundary between the two communes nips off a few houses that by any measure are part of the village of Vougeot. Notwithstanding such hair-splitting, the wines of Domaine Hudelot-Noëllat can be unequivocally declared as being some of the finest reds being produced in the Côte d'Or today.

Charles van Canneyt, who grew up in Reims, had no thought of making wine in Burgundy until he received a call from his grandfather Alain Hudelot in 2006 asking him if he would like to take over management of the domaine. It was too good an opportunity to turn down, so he enrolled in wine school in Beaune, trained with Jean-Louis Trapet in Gevrey-Chambertin and worked a vintage at Giesen in New Zealand in 2009.

He is a man well worth listening to: 'We are still feeling the benefit of global warming, getting excellent maturity ... There is a benefit for the moment, especially for the reds, whites are trickier because people are looking for acidity ... Today, eight out of ten vintages are good, in the old days it was the opposite ... You have to choose the right moment to pick, for there is rapid ripening in the days prior to harvest.'

Van Canneyt's wines speak eloquently too, and not just the trio of *grands crus*: Richebourg, Romanée Saint-Vivant and Clos de Vougeot. Those are glorious expressions of what Pinot Noir is capable of in the right hands, but the entry level Bourgogne Rouge is also worthy of note. Made from 100 per cent de-stemmed fruit it is lively and mouth-watering yet remarkably refined. It is sourced from 18 parcels amounting to 2.3 hectares on the 'wrong' side of the road, opposite Chambolle-Musigny, and accounts for 30 per cent of the domaine's production. As a *régionale* wine it could not be bettered; it has real style and class and would be worthy of *village* ranking.

De-stemming is largely the order of the day for van Canneyt, save for perhaps 20 per cent whole bunches in the top wines: 'Stems sometimes hide terroir, losing the character of each terroir, I prefer to show the differences.' He particularly likes the 2022 vintage: 'It is a style I like, there is good fruit and purity, the wines have a transparency, so you can feel the differences between terroirs. It is more classic than 2018, '19 and '20.'

Hudelot-Noëllat wines sing pure and true, they are polished and poised, there is no swingeing tannin or raw acidity that you must look past in search of beauty, even in young wines tasted from barrel. And followers of this domaine should now look out for an expanded range thanks to van Canneyt and his wife's purchase of Domaine des Chezeaux, which brings some more *grands crus*, notably Griotte-Chambertin, and other choice holdings into the Hudelot-Noëllat fold.

Try this: Vosne-Romanée 1er cru Aux Malconsorts

Malconsorts sits on the mid-slope, abutting La Tâche on one side and the Nuits-Saint-Georges boundary on the other. Hudelot-Noëllat lays claim to a modest 0.14-hectare holding in what is generally regarded as the best of Vosne's *premiers crus*. Indeed, it is not fanciful to suggest that, in Charles van Canneyt's hands, it punches at *grand cru* level. He manages to conjure captivating flavours that go way beyond the normal

120 CÔTE D'OR

descriptors and invite the use of more poetic ones such as beguiling and ethereal. It is a wine of compelling beauty and, given the minuscule production, is bottled only in magnums.

Château de la Tour

Rue de la Montagne, Clos de Vougeot, F-21640 Vougeot
Tel: +33 3 80 62 86 13

With a holding of 5.5 hectares in Clos de Vougeot, Château de la Tour is the largest owner there by some distance, as you might expect from a domaine headquartered in the *clos*. And, were it not for the formidable bulk of the main château, Château de la Tour itself would attract more attention – it is often the case that it needs to be pointed out before people notice it.

François Labet has managed de la Tour since 1984 and was joined by his son, Edouard, in 2018. His grandfather was Jean Morin, after whom the domaine's prestige bottling, Hommage à Jean Morin, is named. Most such cuvées, whether in Burgundy or elsewhere, involve some sort of selection, only using the fruit from an estate's most ancient vines for instance. Labet goes further. The grapes do come from very old vines, the particularity being that only a single bunch, the one closest to the trunk and hence fed first by the sap, is taken from each vine. Needless to say it is only made when Labet deems the vintage to be of sufficient quality.

At a less exalted level a range of wines is also produced under the Domaine Pierre Labet label, named after François' father. These include a perky and fresh Meursault, Les Tillets, of which Labet managed a yield of 11.4 hectolitres per hectare to produce four barrels in 2023. In red, the Beaune typically includes 60 per cent whole bunches and shows the hand of its maker in a more austere flavour than is usually associated with Beaune. The Gevrey-Chambertin, from three plots including La Justice, is muscular and rich with good depth, but some finesse too.

'There is no recipe from one year to the next,' says Edouard Labet when discussing the use of whole bunches, though 90 per cent is typical for the Clos de Vougeot. He is more likely to talk about choice of barrels. Currently five coopers are used, with a previous preference for Chassin now not so pronounced, Rousseau being more favoured in recent years. Tasting the Clos de Vougeot with Edouard from barrel is a barrel tasting as much as a wine tasting. Precisely the same wine – 'it is the same DNA' – from Chassin tastes lighter, with some liquorice and spice, while from Rousseau it is perfumed with a more perceptible tannic bite. In years to come, as Edouard exercises greater influence, the style may shift slightly in that direction.

Try this: Clos de Vougeot Vieilles Vignes grand cru
The term *vieilles vignes* lacks clear definition and hence can be used to imply some extra gravitas in a wine rather than as a statement of vine age. That is definitely not the case here, the vines for this wine are well into their second century, having been planted in 1910. They lie in the heart of the *clos*, almost dead centre on both the north–south and east–west axes. Lush fruit and mild spice define the wine; there is substance and structure but the smooth texture keeps it harmonious right through to the finish. Look for the label with gold writing and not the regular red and black.

VOSNE-ROMANÉE AND FLAGEY-ECHÉZEAUX

Vosne-Romanée might be rural but it is not rustic. There's a reserved feel to this, the most celebrated of all the Côte d'Or villages, a feel given visible emphasis by the manicured vineyards that rub shoulders with manicured properties. There's nothing ramshackle about Vosne; everything is in its place as if ready to serve as a film set. The domaine names, some blazoned surprisingly large on gable walls, others discreet to the point of invisibility, form a prized roll-call, famed across the wine world.

Vosne is the source of the Côte d'Or's greatest red wines, combining finesse and vigour like no other commune. All sorts of arguments can be put forward for the nobility of a great Chambertin or the harmony of a top Musigny, but Vosne, in the person of an on-song top *grand cru*, goes a step further; there's the muscle of Chambertin and the grace of Musigny and then some. The combination and the resulting extra dimension of flavour can make the others seem less complex. It is utterly beguiling and not easy to pin down – 'sweet incense' is my best attempt. Describing a great Vosne in words is a challenge; where the others are often pigeonholed by gender, Vosne is trickier to categorize. There's an intangible quality, full understanding seems to lie just beyond reach;

122 CÔTE D'OR

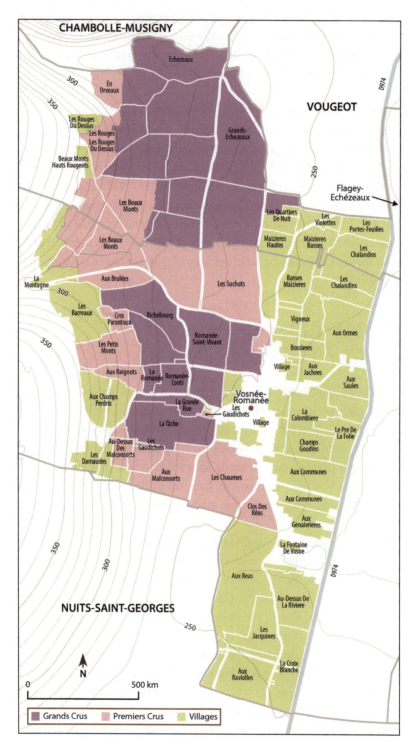

Map 5: Vosnée-Romanée and Flagey-Echézeaux

like poetry, it can be returned to again and again when another shade or nuance of flavour is discovered.

There's treasure in the gentle hillside that rises westwards from the village centre, treasure that wine collectors, behaving more like trophy hunters, are prepared to pay thousands of pounds per bottle for. Bottles of Romanée-Conti and La Tâche are now traded at auction like works of art, which, when considered in the cold light of day, is a ludicrous situation for something that is no more than an agricultural product. As a result the excellent *premiers crus*, at a little less than bank-breaking prices, tend to get overlooked.

Note: Flagey-Echézeaux is included here because its wines are considered, *de facto*, as Vosne-Romanée and include Grands Echézeaux and Echézeaux. Most of the commune land lies on the 'wrong' side of the main road and stretches across the railway line but it is connected to its *grands crus* by an isthmus of vineyard that runs beside the southern boundary of Clos de Vougeot. For hair-splitters the distinction is that the vineyards belong to the commune of Flagey while the wines produced from them qualify for the Vosne-Romanée appellation. The exceptions are the *grands crus*, which are appellations in their own right.

Producers

Domaine Anne Gros
11 rue des Communes, 21700 Vosne-Romanée
www.anne-gros.com; tel: +33 3 80 61 07 95

Anne Gros, who is still very much a presence at her eponymous domaine, has been succeeded by her children, Julie and Paul Tollot-Gros (their father is Jean-Paul Tollot of Domaine Tollot-Beaut). Family interconnections in Burgundy can be labyrinthine, best left to a time-rich genealogist, though for any who are interested the website above contains a complex Gros family tree. It details the links between siblings and cousins, predecessors and successors, links that defy verbal elaboration and which are easier to show in graphic form.

Thanks to those links this domaine recently took over a 0.37-hectare parcel in Grands Echézeaux from Gros Frère et Soeur, occasioned by a redistribution of the Gros family holdings, the first vintage being 2022. 'We don't really know this vineyard yet,' says Paul Tollot-Gros. The *grands crus* they do know comprise the trio: Richebourg, Clos de Vougeot and Echézeaux. The first is undoubtedly the flagship, a wine of great structure and depth, the second is noted as a particularly fine example from the historic *clos*, while the Echézeaux is of greater interest today, thanks to other family inheritances. Paul Tollot-Gros has come into possession, in his own right, of a small plot in the *lieu-dit* Les Beaux Monts Bas, alongside his mother's holding in Les Loächausses. Hence he is now making wine in his own name, strictly speaking as a négociant, 'for the bureaucracy,' as he puts it. (Two domaines cannot operate from the same address. 'I am a négoce but not a négoce,' says Tollot-Gros.)

A little generational rivalry is played out as the two Echézeaux are tasted side-by-side. Tollot-Gros' version echoes his stated aim of wanting to make an Echézeaux, 'with more body.' It utilizes one-third whole bunches to deliver a wine that is plump and rich, with a spicy snap. Meanwhile, his mother's is fresher and more elegant, with a pure line of flavour through to the long finish. Based on comparing the two from the 2022 vintage Anne Gros currently has the upper hand, yet Tollot-Gros quite reasonably asks: 'If I do the same as my mother then why do it?'

Notwithstanding the fun to be had with such comparisons, it would also be foolish for consumers to get fixated only on the top wines of this domaine, for Anne Gros, more than many, has built an enviable reputation for wines from the Hautes Côte de Nuits, most particularly the Cuvée Marine Blanc. First produced in 1999 it combines bright citrus with some richer tropical nuances to give broad appeal. More importantly, and although now more than a quarter of a century old, it points to the future, suggesting that if global warming continues apace the Hautes Côte may yet rival the Côte d'Or for wines of freshness and delicacy. Rival perhaps, though nobody would welcome the great Côte d'Or wines being eclipsed by their upslope cousins. In this, Domaine Anne Gros is ahead of the curve.

Currently, this is a domaine in flux and a gradual change in the house style may occur in the coming years. Long-term followers need not worry, however, quality looks set to remain high.

Try this: Vosne-Romanée Les Barreaux

Barreaux is one of the highest vineyards of Vosne, at about 300 metres elevation, lying upslope of Richebourg and the famed Cros Parantoux. The situation suggests it might have been favoured with *premier cru* status but its largely northern exposure militated against that. In time, with global warming, this may well be its trump card. Thanks to the acidity, the wine delivers a pleasant snap on the palate, together with a spicy, mineral edge, all woven into the dark fruit. It needs time in the cellar to show at its best; then serve it with richly sauced game as the perfect foil.

Domaine Gros Frère et Soeur
6 rue des Grands Crus, 21700 Vosne-Romanée
Tel: +33 3 80 61 12 43

Unlike some domaines, whose premises are discreet to the point of near-invisibility, Domaine Gros Frère et Soeur announces itself by way of a grand building that would not look out of place in central Paris. The memorable aspects do not stop there – a section of the cellar could double as a nightclub with its multi-coloured lighting and a floor that could be used for a remake of *Saturday Night Fever*. A Steinway grand piano in the corner adds a classical touch.

Some of the cellar is brick built and arched, other parts are hewn out of the limestone – visitors walk along a quarried corridor to the elevator that takes them to the penthouse tasting room, completed in 2022. It provides panoramic views over the Romanée-Saint-Vivant vineyard in the foreground, and on across the other Vosne-Romanée

Grand piano and 'disco' floor in the Domaine Gros Frère et Soeur cellars

126 CÔTE D'OR

grands crus, with the domaine's parcel in Richebourg just about visible in the distance.

Compared to the cellar, the wines present a more conventional face, led by the Richebourg, which comes from a 0.6-hectare plot in the Les Vérroilles *lieu-dit*. The vines face east on a noticeable slope, yielding a wine with a gentle nose and a dazzling array of tingling flavours on the palate, without the deep structure more normally associated with Richebourg. Crisp rather than rich, it delivers a lighter iteration of this famed *grand cru*.

Until recently Gros Frère et Soeur produced a quartet of *grands crus* – the Richebourg joined by Echézeaux, Clos de Vougeot and Grands Echézeaux – but thanks to complex inheritance laws this latter holding has now passed to another branch of the Gros family, the 2021 vintage being the *dernière récolte*. The Echézeaux comes from the *lieu-dit* Les Loächausses (all of which is owned by different members of the extended Gros family). A sumptuous nose presages lovely savour and grip on the palate, with marked acidity leading to a clean finish.

The Gros family, along with others such as Colin in Chassagne-Montrachet and Bouzereau in Meursault, traces a complicated lineage. In brief, Gros Frère et Soeur dates from 1963 when siblings Colette and Gustave combined their inheritance from their father, Louis. Bernard Gros, Gustave's nephew, ran the domaine until recently and has now been succeeded by his son, Vincent.

The *grands crus* may garner most attention here but the Bourgogne Hautes Côtes de Nuits in white and red should not be ignored. Both are planted at 10,000 vines per hectare and are exemplary wines, vividly flavoured with impressive length. Seek them out.

Try this: Clos de Vougeot grand cru 'Musigni'

Your eyes do not deceive you – the spelling 'Musigni' is correct and refers to one of the 16 unofficial, and seldom used, *lieux-dits* within the *clos*, this one situated beside the château and closest to the 'real' Musigny across the road. (Others include Quartier des Marei Haut and Montiotes Basses, little wonder they are not used.) With a parcel of 1.5 hectares, Gros Frère et Soeur is one of the largest owners in the *clos* and the wine is marked by a perfumed nose and a light flavour register on the palate. It is only on the finish that its *grand cru* gravitas makes itself felt.

Domaine Nicole Lamarche

9 rue des Communes, 21700 Vosne-Romanée

Tel: +33 3 80 61 07 94

The pace of change in Burgundy may appear glacial but when it happens it can be dramatic, again, like a glacier. Such is the case at 9 rue des Communes, where the cousins Nicole and Nathalie Lamarche managed this domaine jointly until Nathalie removed her share of the vineyards and leased them to Domaine du Comte Liger-Belair (see p. 128). The final vintage from the original domaine was 2021.

'It's not the same but it is good too,' Nicole says, managing a smile, when asked about the loss of 3.5 hectares of *premier* and *grand cru* land. It's a pragmatic response and, in truth, she still has custodianship of a choice roster of vineyards. The loss has been somewhat ameliorated by the addition of bought-in grapes to produce a Nuits-Saint-Georges of real style with a flavour that builds and builds to an exuberant finish, and a Gevrey-Chambertin with sherbet-like vivacity on the palate and memorable intensity on the finish. When asked, in 2024, if she has long-term contracts for these grapes, Nicole replies: 'Five years is agreed for now, there is nothing signed, I cross my fingers …'.

Closer to home, the Vosne-Romanée is delicate and juicy, an exemplar of the house style: 'I am recognized for this style today, showing the elegance and purity of Pinot Noir,' says Lamarche and when asked, for instance, about her use of whole bunches she responds with a non-committal, 'It depends … I always aim for elegance and finesse and perfect balance. There are no rules, it is a question of feeling. I adapt to each vintage.'

The Vosne-Romanée 1er cru Les Chaumes takes things up a notch with firmer tannins and more pronounced acidity and then Les Suchots, one of Vosne's top *premiers crus*, moves up several more notches almost to *grand cru* quality; hardly surprising given that it counts Richebourg and Romanée-Saint-Vivant as neighbours to the south, and Grands Echézeaux and Echézeaux to the north. It is a magnificent wine, ample in every dimension, flavour-packed but with no lack of grace; substance and delicacy interact beautifully all the way to the long finish.

After that, the pair of *grands crus*, Echézeaux and Clos de Vougeot, manage – just – to deliver an extra nuance of flavour: subtle and understated in the former, more pronounced in the latter. They are housed in

128 CÔTE D'OR

a new cellar, constructed in 2024, and also incorporating office space and a tasting room. Despite recent upheaval the future at Domaine Lamarche looks secure for now.

Try this: La Grande Rue grand cru

Three wines vie for inclusion here, the others being the Echézeaux and the Les Suchots. The argument in favour of this *monopole* is incontrovertible, however. It might be regarded as the forgotten *grand cru* of Vosne, having only been promoted to the top rank in 1992 (retroactive to the 1991 vintage), but it keeps the best company, lying sandwiched between Romanée-Conti and La Tâche while also rubbing shoulders with La Romanée and Romanée Saint-Vivant. The wine is an essay in harmony and delicacy, a vinous ballet dancer if such a thing could exist. Tingling red fruits dance on the palate, with all the elements interlaced in a satin texture, allowing the flavour to unfold in the throat long after the wine is gone.

Domaine du Comte Liger-Belair

Château de Vosne-Romanée, 21700 Vosne-Romanée
www.liger-belair.fr; tel: +33 3 80 62 13 70

'Liger-Belair has achieved much since 2000 yet with a new winery planned at time of writing he is not about to coast on those achievements.' With those words I concluded my profile of this domaine in 2017. The winery came on stream in 2019 and Louis-Michel Liger-Belair hasn't coasted either, nor can he, for his domaine has effectively increased in size by 40 per cent since then.

This came about when the two cousins at nearby Domaine Lamarche, Nicole and Nathalie, went their separate ways and divided their vineyard holdings between them. Some top parcels came Liger-Belair's way in 2022 when he agreed a long-term lease with Nathalie, having initially advised her to sell. It will run until 2050 and sees him completely in charge of all vineyard work, vinification and subsequent commercialization of the wines; in all but name he owns the vineyards. They include a mouth-watering array of *grands* and *premiers crus* in good measure and not just a morsel here and there: about half-a-hectare in Clos de Vougeot and Vosne Malconsorts, and a quarter or thereabouts in Grand Echézeaux, Echézeaux, Vosne Les Suchots and Vosne Croix Rameau. Bounty doesn't come much better.

Louis-Michel Liger-Belair explains how it is done – while supported by a barrel of La Romanée

Lamarche can rest easy, safe in the knowledge that, come mid-century, should her children show an interest in the vineyards they will be in impeccable condition, for Liger-Belair is one of the most diligent of Côte d'Or vignerons. He belongs to that small cohort who somehow take their winemaking to a higher level by way of what can only be described as vinous sorcery. Mere words come up short when considering his wines. 'Beauty made wine,' is my best effort; there is poetry in these bottles.

The domaine takes its name from Louis Liger-Belair, a general in Napoléon's army who fought with distinction across the battlefields of Europe in the early years of the nineteenth century. He bought the impressive château in 1815 and over the course of the following decades the domaine's holdings expanded considerably, mainly thanks to astute marriage. At one point the Liger-Belair vineyards read like a roll-call of the Côte de Nuits' best: La Romanée, La Tâche and La Grande Rue held as *monopoles* in addition to sizeable chunks of Chambertin and Clos de Vougeot, as well as a clutch of top *premiers crus*.

It came to an abrupt end on Thursday 31 August 1933 when the vineyards were sold off at auction. Complicated inheritance rules had dictated the sale, because two of the ten children who inherited on the death of their mother in 1931 were underage. Two of the inheritors managed to

130   CÔTE D'OR

buy back La Romanée and some other small plots, which were then rented out for decades, and it is from those embers that the current Vicomte Louis-Michel Liger-Belair has brought the domaine back to life.

And what life. Tasting in the cellars beneath the château is both a rewarding and a frustrating exercise. Rewarding because of the sheer quality of the wines, and frustrating because capturing them in words is like pinning a butterfly to a board, its beauty can only be half-appreciated. Across the range the flavours are bountiful but not oppressive, from the Vosne-Romanée La Colombière all the way up to the jewel in the crown, La Romanée.

Between those poles comes a series of Vosne-Romanée *premiers crus* of rare distinction, including: Les Petits Monts, La Croix Rameau and Les Suchots. The first oscillates beautifully between full and fine flavours; the second is an exotic brew of opulent fruit and fresh acidity; the third leaves them both in the shade courtesy of remarkable intensity and a texture of robust silk.

The Echézeaux takes things a step further in a rolling wave of flavour and then, at the domaine's apex, comes La Romanée. At 0.85 hectare it is the côte's smallest *grand cru* and it lies upslope from and contiguous with Romanée-Conti. There is no wall between them but the boundary is easily spotted thanks to the vines in La Romanée being planted in the relatively unusual north–south direction, compared to the more usual east–west in Romanée-Conti. About half the vines are 100 years old, with another 30 per cent aged 70, and the remainder more recently planted. A parade of enchanting flavours, unfolding seemingly to infinity, defines La Romanée. It is a glorious gustatory aria, deceptively delicate, complete and seamless.

The year 2025 sees the silver anniversary of Louis-Michel Liger-Belair's decision to re-establish his family domaine after a break of many decades. And a bicentenary beckons in 2026 when Liger-Belair will celebrate 200 years' ownership of La Romanée. History's warp and weft has been cruel and kind to this domaine; it's all the latter right now.

Try this: Vosne-Romanée 1er cru Aux Malconsorts
The nearly 6-hectare Malconsorts is bordered by La Tâche to the north and the commune boundary with Nuits-Saint-Georges to the south. Liger-Belair's half hectare is part of the Lamarche 'bounty' and is a wine of extraordinary quality that displays a carousel of flavours, spiralling across the palate and defying the normal prosaic descriptors to capture

its beauty. It is exotic and heady, lush yet fresh, full yet fine. All the components are there in ample measure, harmonious and balanced, and echoing long into the reverberating finish.

Domaine Méo-Camuzet

11 rue des Grands Crus, 21700 Vosne-Romanée
www.meo-camuzet.com; tel: +33 3 80 61 55 55

Domaine Méo-Camuzet warrants the highest approbation, if only because of its excellent Clos de Vougeot and Corton, whose reputations could use some burnishing. This pair are exemplars and should be championed as such, held up as *vrais grands crus* able to stand compare with their smaller, more celebrated neighbours, particularly the Vougeot, which rubs shoulders with Musigny and Grands Echézeaux. The domaine is one of the largest owners in the *clos*, with a holding of three hectares, the main part of which is located beside the château itself. From soil that is only 40 centimetres deep comes a wine of poise and conviction, there's amplitude and depth, oodles of fruit enlivened by sweet spice, and a tingling departure on the finish.

It is appropriate that Méo-Camuzet should produce a nonpareil Clos de Vougeot, for the domaine's history is intertwined with it. Etienne Camuzet (1867–1946) served as mayor of Vosne-Romanée and deputy for the Côte d'Or from 1902 to 1932. More pertinently, he bought the Château du Clos de Vougeot and 3 hectares of vines in 1920, later donating the château to the Confrérie des Chevaliers du Tastevin. For any whose enthusiasm for Clos de Vougeot is no better than lukewarm Méo-Camuzet's will restore confidence thanks to a remarkable kaleidoscope of flavours.

After Camuzet's death the domaine passed to his daughter and then to her nephew, Jean Méo, who was succeeded by current incumbent Jean-Nicolas Méo in 1989. He cultivates the vineyards organically though has never sought certification, and always de-stems 100 per cent of the fruit. Thereafter the grapes are cooled for a few days' maceration and, during fermentation in concrete vats, two short pump-overs per day are employed early on with some punching down towards the end. For the élevage, the top wines, such as the Richebourg, see 100 per cent new barrels, with 80 per cent for the other *grands crus*, 60 for the *premiers crus*, and none for the Fixin, Marsannay and Bourgogne Rouge.

Richebourg is the stand-out wine. After a deceptively light entry a remarkable cornucopia of flavours gradually unfolds, oscillating between lean and rich elements with no conflict between them. The Méo-Camuzet holding amounts to one-third of a hectare, comprising two parcels, the majority lying at the top of Les Vérroilles, augmented by little more than a scrap in Richebourg itself.

Amongst the domaine's *premier cru* holdings, three-quarters of a hectare in Vosne-Romanée Aux Brulées produces a wine that could challenge many *grands crus* and is treated as such by Méo-Camuzet. It may sound pejorative to describe it as 'easy drinking' but that is exactly what it is, for here is a wine of harmony and elegance, with a texture that caresses the palate as it passes across. The Méo-Camuzet website states: 'The objective of the estate is to produce wines that combine structure and finesse, concentration and charm.' Jean-Nicolas Méo and team are nailing that objective.

Try this: Corton Clos Rognet grand cru

This walled vineyard of 0.4 hectares plays host to the domaine's oldest vines, planted in 1927, and lies in the northern part of Corton with an exposure directly to the east. The wine is notably structured with a firm bedrock of tannin counterpoised by abundant fruit. The lasting impression is of a vivid, intense, energetic mouthful. Any wine lovers who never look beyond the Côte de Nuits for truly great red burgundy should try this.

Domaine Georges Mugneret-Gibourg

5 rue des Communes, 21700 Vosne-Romanée
www.mugneret-gibourg.com; tel: +33 3 80 61 01 57

There has been a generational shift at Mugneret-Gibourg since the first edition of this book. Matriarch, Jacqueline, who took over the running of the domaine after the early death of her husband, Georges, in 1988, died in 2024. Her daughters, Marie-Christine and Marie-Andrée, who worked with their mother for decades, have now been joined by Marie-Christine's daughter, Lucie, and Marie-Andrée's daughter, Marion. Fanny, Marion's sister, manages the guesthouse next to the domaine, named for her grandmother, La Maison de Jacqueline.

Georges Mugneret's principal occupation was as an ophthalmologist in Dijon but his true interest lay in the small domaine started by his parents in 1933, which he enlarged considerably in the decades following his first purchase, in 1953, of a 0.34-hectare parcel in Clos de Vougeot.

Thanks to this and other astute purchases the domaine, though based in Vosne-Romanée, now has holdings scattered the length of the Côte de Nuits, from Gevrey-Chambertin to Nuits-Saint-Georges. This is partly thanks to the influence of Charles Rousseau of Domaine Armand Rousseau, who was a good friend of Georges Mugneret. Discovering in 1977 that a large parcel of *grand cru* Ruchottes-Chambertin was for sale he urged Mugneret to buy a portion of it and when the latter demurred, saying he didn't know Ruchottes, Rousseau silenced his qualms with a simple 'follow me'. The purchase of a 0.6-hectare parcel went ahead and today Ruchottes is the Mugnerets' flagship wine, outstripping the Echézeaux and Clos de Vougeot in every dimension of flavour.

More recently, in 2018, an extension to the cellar was added to accommodate an extra two hectares worth of production, when a share-cropping agreement came to an end. It is already well blanketed in dark mould, making it appear aged beyond its years.

The grapes arrive at the winery in 22-kilogram boxes, where 100 per cent de-stemming is largely the norm, save for the Echézeaux and the Ruchottes which utilize about 10 and 20 per cent respectively. Fermentation is in concrete vats, with a combination of pumping over and punching down – daily tasting dictates which. Trade and press visitors with pencils poised above notebooks, or fingers ready to tap on keyboards, awaiting precise facts and figures in answer to their questions, will come away frustrated: 'We are not very technical winemakers … we cool the grapes a little … I am scientific but the magic part of our wine is to work with the conditions of the year … there is no recipe. It's like when I cook,' Marie-Andrée explains. The only thing that can be stated definitively is that the entire production is red: 'I have never tried to make white wine from Pinot Noir,' she quips.

Try this: Vosne-Romanée
The wine is assembled from a quintet of *village* vineyards stretching from Les Chalandins at the north of the commune to La Croix Blanche at the south, with a trio in the middle: Les Champs Goudins, La Colombière and Le Pré de la Folie – the last of which is translated by Marie-Andrée as, 'the field of madness.' There may be madness in the field but not in the glass, for this is quintessential Vosne, with the indefinable extra that sets it apart. The nose is fragrant and inviting, while the cherry fruit on the palate is shot through with a streak of sweet incense.

Domaine Georges Noëllat

1 rue des Chaumes, 21700 Vosne-Romanée

Tel: +33 3 80 61 11 03

Apart from the obvious family link between this domaine and Hudelot-Noëllat in Vougeot, a more obscure one stems from the fact that the men in charge at each domaine – Maxime Cheurlin here and Charles van Canneyt at Hudelot-Noëllat – both grew up in Champagne before answering the call of Burgundy. Cheurlin arrived here in 2010 to take over from his grandmother and immediately made his mark, courtesy of wines that displayed all the elegance and conviction one can expect from Vosne-Romanée.

Indeed, even the non-Vosne wines such as the Beaune 1er cru Clos de la Mignotte and the Chambolle-Musigny 1er cru Les Feusselottes, seem to carry a Vosne overlay, like a person acquiring the accent of their adopted country. The house style dovetails most successfully with wines closer to home, such as the pair of *premiers crus* Les Beaux Monts and Les Petits Monts, near neighbours on the upper slopes above the village of Vosne-Romanée. The former makes an impression through finesse, not power, with fine acidity and dark fruit; the latter sweet and soft with oodles of cherry and strawberry fruit threaded with crisp tannin.

The style reaches its apogee in the Grands Echézeaux and the Echézeaux, where the domaine's holdings are modest: a thin slice of 0.3 hectares that runs the width of the vineyard in Grands Echézeaux, and barely more than one-tenth of a hectare at the extreme southern end of Echézeaux. Abundant, succulent, rich fruit, with a touch of liquorice adding opulence, defines the Grands Echézeaux, while the Echézeaux displays a lighter weave, beautifully fine and fresh with discreet tannins.

Vinification starts with a, roughly, 100 per cent de-stemming regime, though the Echézeaux, for instance, will see 30 per cent whole bunches. Then follows a cool maceration at 5°C for one night, and fermentation of about three weeks. Of particular interest at Georges Noëllat is the oak regime for the élevage. The barrels all come from the Cavin *tonnellerie* – 'Aphrodite' being the model favoured by Cheurlin – and are given a long and light toasting. Each carries precise details of its manufacture, almost like a passport: the model, the year of production, the level of toast, the length of ageing of the staves, the forest the oak came from, the barrel number, and a code for traceability. The *grands crus* see 100 per cent new barrels, diminishing to 50 for the *premiers crus* and 30 for the *village* wines. Across the range these are wines of style, and increasing grandeur, as you move up through the appellations.

Try this: Vosne-Romanée 1er cru Les Chaumes

Depending on your standpoint this vineyard could be considered the domaine's front or back garden, though Noëllat only lays claim to about 0.5 hectares out of the total 6.5. Only a narrow road separates Les Chaumes from La Tâche at its north-west corner and, although it cannot claim the same exalted status, this wine is not shamed by comparison, thanks to rich but reserved fruit and an overall impression of delicate intensity that carries into the lingering finish.

Domaine Michel Noëllat

5 rue de la Fontaine, 21700 Vosne-Romanée

www.domaine-michel-noellat.com; tel: +33 3 80 61 36 87

Domaine Michel Noëllat takes its name from the grandfather of the current generation, cousins Sophie and Sébastien, and has not enjoyed quite that same renown as some of its more celebrated neighbours in Vosne-Romanée. Under their stewardship, however, the domaine is in a state of positive flux and the wines are now well worthy of their place in Vosne's firmament. The domaine comprises 25 hectares spread across 23 appellations, divided into 100 separate plots. Some Côte de Beaune wines are made but this is principally a Côte de Nuits estate, producing red wines from all the côte's villages except Gevrey-Chambertin.

For those wines a primary selection of the fruit is made at the vineyard, then another at the winery, where the grapes are 100 per cent destemmed. 'We don't use any whole bunches, we like the fruit,' explains Sophie Noëllat, while also admitting to a liking for whole cluster wines from other domaines. There then follows a cool maceration at 10°C for five or six days before fermentation, during which *remontage* is favoured over *pigeage*, though there might be some manual *pigeage* towards the end. After fermentation the wine is left to rest for several days' infusion before transfer to barrel, principally from François Frères, Cavin and Taransaud. Standard *pièces* are used for the principal wines, with 500-litre barrels for the humbler appellations.

136 CÔTE D'OR

The wines have an immediate sensual appeal, from beautiful perfumes to svelte palates. The basic Hautes-Côtes de Nuits is an exemplar for the house style, a remarkable wine that punches far above its station, juicy and fresh, pure as birdsong. Moving up through the appellations, the Chambolle-Musigny Les Fouchères is made from four plots of 60- to 80-year-old vines, whose small berries bestow superb concentration with no loss of Chambolle's signature finesse. The Vosne-Romanée Aux Genaivrières was made for the first time in 2022 and comes from a vineyard that is barely 100 metres from the winery. It is a sure-footed wine with grippy tannins that bode good future development.

The Nuits-Saint-Georges 1er cru Aux Boudots is plump and substantial, 'the perfect winter wine with fire and snow,' according to Sophie Noëllat. Then, vying for top spot, are the domaine's pair of *grands crus*: Clos de Vougeot and Echézeaux. The former comes from a holding of 0.4 hectares in two plots and is a dark, brooding wine that takes time to unfold and reveal the mild savour and perky spice that add complexity to the succulence of the house style. The latter comes from the *lieu-dit* du Dessus and sees 16 months of élevage in barrel, 70 per cent of it new. The fruit is dark and the texture is dense, yet the wine is not overbearing thanks to a lighter twist on the finish.

This is a name to watch, not only for the wines of Michel Noëllat but also because, at time of writing, Sophie and her husband Arnaud Sirugue were building a new winery to produce wines in their own name, Sirugue-Noëllat.

Try this: Vosne-Romanée 1er cru Les Suchots
A glance at the map suggests that Les Suchots, bracketed by Grands Echézeaux to the north and Romanée Saint-Vivant to the south, might have been accorded *grand cru* status when such things were being decided. Only when you place your feet on the ground do you discern the slight declivity in which Suchots sits, robbing it of the highest rank. Nevertheless, this is a compelling wine and Noëllat, with a holding of 1.8 hectares across two plots, produces it in good quantity. Beautiful balance means that the firm structure is barely discernible behind its cloak of sweet fruit, leading to a clean and fresh finish.

Note: The domaine suggests a serving temperature of 13–15°C, which is well below today's 'room temperature'.

Domaine Armelle et Bernard Rion

8 route Nationale, 21700 Vosne-Romanée
www.domainerion.fr; tel: +33 3 80 61 05 31

This domaine takes full advantage of its easy-to-spot location on the D974 opposite the village of Vosne-Romanée, being open for cellar door sales, including Burgundian truffles, and regularly welcoming tour groups from Viking River Cruises (who arrive by bus). Oenotourism plays an important role at Rion and that may point the way forward for other small domaines, so long as it doesn't compromise the winemaking. Armelle and Bernard Rion have now handed over to their three daughters, Alice, Nelly and Melissa, with Alice in charge of winemaking and her sisters dividing the other responsibilities between them. Vineyard holdings amount to 9.5 hectares, supplemented by a small négociant business.

Alice's husband, Louis, is in charge of viticulture and keeps vineyard treatments to a minimum, primarily for the vines but also not to interfere with the truffle growth. Pruning is now carried out later than previously to try and mitigate the threat of spring frost, and harvest is by hand. In the winery the grapes are 100 per cent de-stemmed before fermentation in relatively new concrete vats that date from 2019. A combination of punching down and pumping over is utilized during the 12- to 15-day process, before élevage in barrels from Montgillard, Meyrieux, Claude Gillet and Rousseau (made with wood from the nearby forest of Cîteaux). Standard 228-litre *pièces* are used for the reds and larger 450-litre *demi-muids* for the whites.

Use of new oak rises from 10 per cent for the regional wines to roughly 50 for the *premiers crus*, while the Clos de Vougeot is treated to 100 per cent. The Rion holding of 0.16 hectares is the second smallest in the *clos* and comprises five long rows running from the road into the centre of the vineyard, yielding enough to make three barrels.

Other wines include a safe and sound, rather than special, rosé blended from Pinot Noir and Gamay; a perfumed and crisp Bourgogne Rouge La Croix Blanche from a *lieu-dit* across the road and barely 100 metres from the winery; and a well-structured old-vine Vosne-Romanée 'Dame Juliette', named for the current generation's grandmother. There is also a fresh and racy Chambolle-Musigny from a *lieu-dit* above the village with the bound-to-confuse name, Les Echéseaux (also spelt Echézeaux).

138 CÔTE D'OR

Try this: Nuits-Saint-Georges 1er cru Les Damodes Vieilles Vignes
Les Damodes lies at the northern limit of Nuits, on the boundary with
Vosne-Romanée and abutting Vosne's Les Damaudes. Whatever the
spelling, the name may derive from *les dames hautes*, the high ladies,
possibly referencing the vineyard's altitude, rising to 350 metres at its
highest point. It faces east with a slight turn north and that, combined
with the altitude and the proximity to Vosne, results in a wine far superior to all the other Rion wines, save the Clos de Vougeot. It is fine
and elegant, a wine of purity and precision with deep reserves of flavour
that continue into the satisfying finish.

Domaine de la Romanée-Conti

1 place de l'Eglise, 21700 Vosne-Romanée

www.romanee-conti.fr; tel: +33 3 80 62 48 80

Domaine de la Romanée-Conti, usually referred to as 'DRC' by wine
lovers across the globe or 'the Domaine' by some closer to home, is the
most famous wine-producing estate on earth. And yet – when one visits
the discreet premises beside the church in Vosne-Romanée, the impression is of an extraordinary disconnect between image and reality. The
image derives from the frenzied atmosphere and faux sophistication of
the auction room, as bidders chase after precious bottles from celebrated
vintages. Ally to that the fawning ambience and theatrical ceremony in
high-end restaurants when an old bottle is broached and you have an
effective smokescreen that all but obliterates the fact that this is an agricultural product made by farmers. The reality is more prosaic, initially
signalled by the name in tiny lettering on the bell at the entrance gate;
eye-catching brass plaques are for others. The reception area is well appointed though in no way lavish, superlatives can be saved for the wines.

The domaine's centrepiece, the 1.8-hectare Romanée-Conti vineyard, has been celebrated for centuries and on the few occasions when
it changed hands it commanded fabulous prices. It was first called La
Romanée by the de Croonembourg family who owned it in the sev-
enteenth and eighteenth centuries before selling to Louis-François de
Bourbon, Prince de Conti in 1760. He added his name while keeping
all the wine for himself and it has remained 'Romanée-Conti' since.
After the Revolution it was seized by the state and sold off as a *bien
national*, going through a succession of owners until Jacques-Marie
Duvault-Blochet bought it in 1869. The domaine as constituted today
owes its origins to him.

Edmond de Villaine married into the Duvault-Blochet family in 1906 and for some 40 years co-owned the domaine with his brother-in-law Jacques Chambon, until the latter needed to sell in 1942. His share was bought by Henri Leroy, a négociant from Auxey-Duresses. Since then it has been jointly owned by the de Villaine and Leroy/Roch families and until a few years ago it was co-managed by Henri-Frédéric Roch and Aubert de Villaine, a direct descendant of Duvault-Blochet. Today, they have been succeeded by Perrine Fenal, Henri Leroy's granddaughter, and Bertrand de Villaine, Aubert's cousin.

For many years Aubert de Villaine was the public face of DRC, and he remains a quiet presence there still. Hearing him talk of his earliest memories makes for a reality check. His father, Henri Gaudin de Villaine, was essentially a cattle farmer who visited Vosne-Romanée once a week in his role as manager to supervise vineyard and cellar work. More tellingly, he also used earnings from his farming to subsidize wages at the loss-making domaine. As Aubert tells it: 'The domaine, since the phylloxera arrived in Burgundy in the 1870s and 1880s until 1972, never made a cent of profit ... In fact, the proprietors at the time had other professions in order to make a living.' Times have changed radically since then and as Bertrand de Villaine conducts a barrel tasting, drawing samples – and subsequently pouring back any unfinished drops of precious wine – it is hard to believe that it was once difficult to sell these wines. He is cut from different cloth to Aubert, bursting with bonhomie, where Aubert is more reserved.

'Don't tell me what you put in your cellar, tell me what you put in your stomach!' Bertrand de Villaine doesn't mince his words when expressing his disdain for those well-resourced wine collectors who love to enumerate their expanding collection of wine while appearing never to drink any. 'It is not a crime to drink young wine,' he continues, agreeing that stocking a cellar full of vinous Picassos and Rembrandts is a sterile exercise, 'A wine lover is curious, he likes to taste and drink his wines.'

Some gradual change has been evident at DRC in recent years, notably the move to using 100 per cent whole bunches since 2017, where 80 per cent had been the norm previously. Bertrand quips that the de-stemmer is now in the *grenier* (attic), adding, 'we will use it if we want.' The change has added some structure without markedly altering the style. Fermentation is in large oak vats and lasts about 20 days before pressing and racking into barrel, all François Frères. Élevage takes place in a series of interconnected cellars into which air-conditioning was installed in summer 2024. 'We don't need it yet but we are planning ahead,' says de Villaine, adding, 'it is a big decision and maybe we will never use it, but it is like having a fast car. You don't have to drive at top speed but it is nice to know you can if you need to overtake a truck.' There is certainly no intention of holding the temperature rigidly constant year round: 'Life is not 12°C all the year.'

Tasting from barrel involves a pleasant amble about the cellars, seeking out samples in ascending order of renown. The ripe and abundant 2023 vintage, tasted in July 2024, was generally open and approachable, particularly the plump and succulent Echézeaux with its leavening of spice and liquorice, for which my tasting note concluded: 'Could drink it now.' By contrast, the Grands Echézeaux was less forthcoming and, in a reversal of the normally accepted hierarchy, it was outshone by its nominally lesser neighbour. In both of these vineyards DRC is comfortably the largest owner, with holdings that dwarf almost all others.

There is always a considerable leap in class when moving from the two Echézeaux to the next pair, the Romanée-Saint-Vivant and the Richebourg, which lie opposite one another across a narrow road.

Tasting at Domaine de la Romanée-Conti

Again, DRC is the major landholder in each, with 5.29 hectares in the former and 3.51 in the latter. Commentators often cast the pair as rivals, wondering in any given vintage if the finesse of the Saint-Vivant will outshine the power of the Richebourg, or vice versa. I normally favour the Richebourg but in 2023 there was no contest, the Saint-Vivant had an indefinable polish that set it apart. It was graceful, refined, elegant and delicious thanks to a flow of wonderful fruit that seems endless. 'This is what I love in burgundy wine, my queen,' said de Villaine. By contrast the Richebourg was closed and commanding, standing back where the Saint-Vivant came forward, its hidden power only hinted at.

The two *monopoles*, Romanée-Conti and La Tâche, were richly endowed in every respect. The former is usually rated highest but perhaps that is because La Tâche, at six hectares, is considerably larger and so not as rare. It lies a little south of Romanée-Conti and is separated from it by the vineyard sliver of La Grande Rue. La Tâche is always a favourite and the 2023 did not disappoint. It was the most charming, most beguiling of them all. Repeated tasting revealed facet after facet of flavour; no sooner had one been perceived than another presented itself. A rolling wave, sustained by fruit in remarkable abundance. The Romanée-Conti impressed with power and depth, a sense of increasing flavour momentum as it crossed the palate, layer after layer revealed like opening a Russian doll.

South of Vosne-Romanée the domaine has three outposts in the Côte de Beaune, all *grands crus* as might be expected. Since 2008 the Corton vineyards of the late Prince de Mérode have been leased, the first vintage being 2009 and, further south, three small parcels amounting to two-thirds of a hectare are owned in Le Montrachet. Corton-Charlemagne is the most recent addition (below).

DRC wines are remarkably appealing on release, they are not locked in a fortress of extraction or hidden behind a wall of swingeing tannin, yet they can age beautifully too. Witness an Echézeaux 1964 that was intense and lively when drunk in 2018. Granted, it came straight from the domaine's cellars but a La Tâche 1947 drunk in 2021 certainly did not and was showing all the signs of a rough life – the label was careworn and the fill level was worrying, yet despite its tribulations it blossomed in the glass for 20 minutes before gently declining. A Richebourg 1954 – an unheralded vintage – drunk in 2023, was equally compelling.

142 CÔTE D'OR

Domaine de la Romanée-Conti also produces a tiny quantity of Bâtard-Montrachet from a holding of 0.12 hectares, though it is never released commercially and is only served at the domaine. It might be called the 'house white'.

Try this: Corton-Charlemagne grand cru
Perhaps a surprising choice from a domaine renowned for red wines but this is something new and different and, indubitably, exceptionally good. It is made from four plots with a combined surface of about 3 hectares, leased from Bonneau du Martray, the first vintage of which was 2019. For that maiden vintage and 2020 the plots were vinified separately but subsequent years have seen them vinified together. There is opulence and abundant exotic fruit on the nose, followed by a plump and fleshy palate, held in check by crisp minerality and fresh citrus. The flavour oscillates between richness and reserve, dazzling the palate. In short, the wine challenges the vinous lexicon, stretching it to accommodate such amplitude and generosity of flavour.

NUITS-SAINT-GEORGES & PRÉMEAUX-PRISSEY

The great sweep of *grand cru* vineyard that runs through half a dozen communes from Gevrey to Vosne halts before Nuits and the regular pattern of tiny villages nestled amongst prized vineyards unravels a little, as if presaging the more topographically varied Côte de Beaune to the south. The town itself splits the appellation like a wedge driven into the hillside along the course of the Meuzin river and into the Gorges de la Serrée, between slopes that rise to 400 metres on either side. Though not large, Nuits has the feel of a metropolis compared to, say, Chambolle.

Before acquiring its current name the town initially suffered the indignity of being labelled Nuits-sous-Beaune by the railway company to avoid confusion with Nuits-sous-Ravières, north-west of Dijon. The capital of the Côte de Nuits couldn't possibly be defined by reference to its southern rival so after some wrangling it appended the name of its most prestigious vineyard, Les Saints-Georges. Perhaps it's the 'Saint-Georges' but Nuits has always enjoyed good visibility in the English-speaking world, bolstered by a reputation for solid and dependable wines. That dependability gave the lack of excitement a veneer of

THE VILLAGES AND PRODUCERS OF THE CÔTE DE NUITS 143

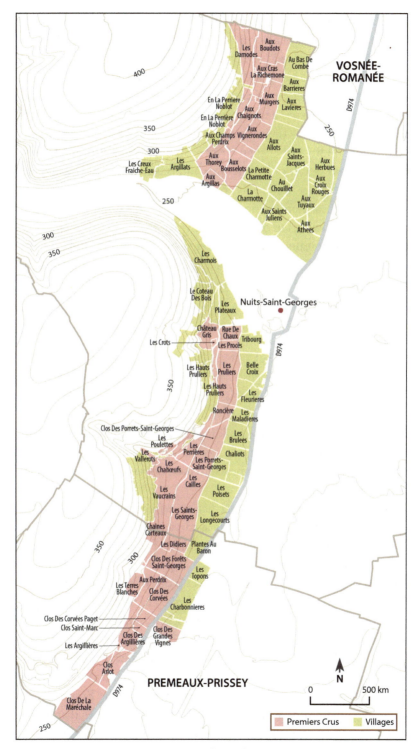

Map 6: Nuits-Saint-Georges

144 CÔTE D'OR

acceptability. Burly flavours were often the hallmark and pulses seldom raced in anticipation of a bottle of Nuits, in the way they might if a fine Chambolle or structured Gevrey was proffered.

Nuits can be strapping and tannic in a way that is rare in the Côte d'Or. Where the strength of a good Gevrey can be supple, that of Nuits can veer towards coarse and only the most dextrous winemaking hands get the best out of it. Their wines can be delicious yet the received reputation for solidity rather than charm still holds sway. The good news, though, is that dexterity in vineyards and cellars is on the rise and the stereotype, once cast in stone, is not as solid as heretofore. 'Elegant', previously an unheard-of descriptor, now appears in tasting notes. South of Nuits a narrowing tail of *premier cru* vineyards crosses the commune border into Prémeaux-Prissey, bracketed by road and hillside. In the same way that the Flagey vineyards qualify for the Vosne appellation so these are included in the Nuits appellation.

Producers

Maison Jean-Claude Boisset

Les Ursulines, 5 quai Dumorey, 21700 Nuits-Saint-Georges

www.boissetcollection.com/brands/jean-claude-boisset; tel: +33 3 80 62 61 61

Though nearly a decade old, Boisset's Les Ursulines winery still has the power to set first-time visitors' jaws dropping and send them reaching for their camera phones. The building has aged well but more importantly it has served exceptionally well as winemaker Grégory Patriat's production facility, though so prosaic a term does no justice to its splendid architecture.

Boisset's origins date back over half a century to 1961 when Jean-Claude Boisset started in business aged eighteen. He quickly proved adept at snapping up ailing producers, some of which dated back to the eighteenth century. The rate of shrewd acquisitions reached breakneck speed in the 1990s, so that today Boisset is one of the biggest wine producers in France. Yet most wine lovers, if asked to list Burgundy's largest producers, will rattle off the names of négociants such as Drouhin, Jadot and Latour. Few will mention Boisset, in terms of size the daddy of them all. Many of the Boisset companies lie outside the scope of this book, such as J Moreau in Chablis, Thorin in Beaujolais and wineries much further afield in Canada and California. Others are part of the Boisset group but are maintained as semi-autonomous operations, rather than being absorbed into one giant Boisset entity, hence keeping it under the radar.

Vat room at JC Boisset winery 'Les Ursulines', Nuits-Saint-Georges

These include Bouchard Aîné, Pierre Ponnelle, Jaffelin, Mommessin, Louis Bouillot, Ropiteau Frères, F Chauvenet, Alex Gambal and Charles Vienot. The Boisset name itself is reserved for the wines produced under the 'Jean-Claude Boisset' label at Nuits-Saint-Georges.

Patriat has been in charge here since 2002, an inspired appointment by Jean-Claude Boisset and his children, Jean-Charles and Nathalie. 'I have no filter,' he says as he leads visitors on a cellar tour enriched by a torrent of lively asides, many memorable, few printable. He hasn't used bâtonnage on his white wines since a conversation with Jean-Marc Roulot in 2007 convinced him it wasn't necessary. 'It is like make-up, it gives an artificial fatness and hides the terroir,' he explains. The wines lose nothing from this approach, they are well composed with pure, precise flavours and notable freshness. For the vinification, whole clusters go straight to the press with no crushing and there is no *debourbage* either, the aim being to get as much of the lees into barrel as possible. Patriat likens this to 'vaccinating the wine' to protect it against premox. He uses 450-litre barrels, 'the magnum of the barrel,' he says, explaining that the staves are 27 millimetres thick and would have to be thicker if the barrels were any bigger, thus affecting the transpiration through the oak. The wines spend two winters in wood before blending in stainless steel and bottling.

146 CÔTE D'OR

Despite the quality of the whites, the reds go a step further, by way of verve and vibrancy; where the whites caress and charm the palate the reds demand greater engagement. Patriat's particular skill is to coax elegant flavours from terroirs that would drift to coarseness if not handled properly, specifically any of his Nuits-Saint-Georges bottlings. Leaving them in the shade, however, is the Vosne-Romanée Les Petits Monts, a *premier cru* of fabulous, restrained power, concentrated yet smooth as it resolves into an echoing finish.

For his reds Patriat generally uses 50 per cent whole cluster, 'too much will obscure the terroir,' vinified in open-topped wooden vats. He starts with a cool maceration at 12–13°C for 5–7 days and only uses natural yeasts. He likes a long fermentation and aims to 'do as little as possible,' emphasizing that 'there is no spreadsheet,' no set of diktats to be followed slavishly. Few winemakers are as well worth listening to as Patriat, and not only can he talk the talk, his wines walk the walk.

Try this: Aloxe-Corton 1er cru Les Valozières

Selecting a single wine for special mention from the dozens in the Boisset range is a challenge. The Criots-Bâtard-Montrachet and the Clos de la Roche are both worthy of the nod but are only produced in minuscule, one-barrel quantities, so the spotlight falls on this *premier cru* that lies downslope of Corton Les Bressandes, facing roughly east. Patriat declared himself 'really proud' of the 2022 vintage, a wine of bountiful fruit with tannin and acidity acting as parameters on the expansive flavour.

Domaine Faiveley

8 rue du Tribourg, 21700 Nuits-Saint-Georges
www.domaine-faiveley.com; tel: +33 3 80 61 04 55

Bicentenary bells were ringing at Domaine Faiveley in 2025 to celebrate the two-hundredth anniversary of the house's founding by Pierre Faiveley in 1825. In the two centuries since, Faiveley has been a fixture in Nuits-Saint-Georges and today the company is in the charge of seventh-generation brother and sister, Erwan and Eve. The family also occupies a unique place in Burgundian history thanks to the dynamism of Georges Faiveley, the siblings' great-grandfather, who founded the Confrérie des Chevaliers du Tastevin in the depressed 1930s to help promote Burgundy. He is well remembered for his adage: 'If no-one wants to buy our wines then let's invite our friends to share them with us.' It was a response to a dire situation that is barely imaginable today.

Vat room at Domaine Faiveley, Nuits-Saint-Georges

A no-less important anniversary in the company's history sees Erwan Faiveley celebrating his twentieth year in charge – and in that time he has proved himself to be much more than a safe pair of hands, guiding a long-established family business along a preordained path. Far from being inhibited by the weight of family history he has taken inspiration from it and directed it along a different path, overseeing the change from négociant to domaine status. For visitors to Nuits this is most visible in the splendid new vat room and, most notably for consumers, in a radical shift in the house style. The forbidding, extracted style of yore that took decades to mature, has given way to greater finesse and elegance; sweet fruit has replaced severe tannins (though the old style could be magnificent when properly mature).

At time of writing, organic certification was in the pipeline, though Eve Faiveley, who joined the business in 2014, suggests that the change in viticulture is not perceptible in the wines: 'It is difficult to say, there is no big difference.' They are unlikely to progress to biodynamics, Erwan being sceptical of its practices. For the red wines, which dominate production, the house policy is to use whole bunches sparingly, with some exceptions, such as the Musigny (below). They are vinified in open-top wooden and stainless steel vats, cooled initially to 8–10°C for four to five days before fermentation, during which a combination

148   CÔTE D'OR

of *remontage* and *pigeage* is used, depending on the style of the vintage. Élevage follows in standard *pièces* from a quintet of coopers: François Frères, Taransaud, Cadus, Meyrieux and Chassin.

Many dozens of different wines are produced, including a wide roster of *premiers* and *grands crus*, though it is not necessary to go to the top of the tree to sample the new Faiveley style. First produced in 2020, the Nuits-Saint-Georges Les Montroziers, a cuvée blended from five *lieux-dits* – Aux Lavières, Tribourg, Les Damodes, Les Argillats and Belle Croix, totalling 6.5 hectares – is accessible and satisfying, its calling card being glossy sweet fruit that comes on in a delicious wave before subsiding to reveal a slightly firmer, spicier element. That said, treasure aplenty awaits at *grand cru* level, most particularly the flagship Corton Clos des Cortons Faiveley, the only *grand cru* to carry the name of its owner, though a similar, if weaker, claim can be made for Romanée-Conti. The 3-hectare vineyard (which includes a small plot of Chardonnay) lies in the Ladoix-Serrigny commune near the top of the slope and bordering Les Renardes. The style is restrained, deep rather than broad, with tingling acidity and smooth tannin supporting the sweet fruit into the lingering finish. Based on wines such as these Faiveley is well set as the business starts into its third century.

Try this: Musigny grand cru

Faiveley lays claim to a modest holding of 0.13 hectares split across two plots at the northern end of this fabled vineyard. The vines were planted in 1945 and 1990, and the wine is made entirely with whole bunches. Production runs to about 400 or 500 bottles a year and the wine is aged in *feuillettes*, half-size barrels of 114-litre capacity, 100 per cent new. The result is an intense and forceful wine, structured and strong, that delivers a panoply of flavours in addition to the generous fruit. There is a herbaceous whiff and a sweet vein of incense-like spice and, it hardly needs saying, an impressively long finish.

Domaine Henri Gouges

7 rue du Moulin, 21700 Nuits-Saint-Georges
www.gouges.com; tel: +33 3 80 61 04 40

If pressed to choose a single domaine that represents the essence of Nuits-Saint-Georges many burgundy lovers would nominate Domaine Henri Gouges. Founded by the eponymous Henri in 1919 it is now in the hands of the fourth generation, cousins Grégory and Antoine

Gouges. For much of that time the house style was a reference for red burgundy of uncompromising stripe, but with the domaine now into its second century modest change is in the air, as the cousins seek to refine the style while remaining true to their heritage. They tread a delicate path, balancing the traditional, forbidding signature with a desire to make more approachable wines, for those wines of old, while appealing to purists, were off-putting to any who weren't prepared to bury them in the cellar for decades, not mere years.

The cousins are doing a fine job – and the change is signalled in the bright new visitor reception area, a marked contrast to what went before. They are working for more finesse and accessibility but also want to retain the capacity for ageing. According to Antoine, the style has been evolving since 2015, by way of greater definition in the wine, 'but I am not worried about the ageing potential, it doesn't come from tannin but from concentration.'

This is a Nuits house through and through; all the holdings lie within the commune boundaries, some 14 hectares in all that include a sextet of *premiers crus* plus 3 hectares of *village* and a small amount of white. Save for two small cuvées the grapes are 100 per cent de-stemmed, followed by a gentle maceration of intact berries and minimal use of sulphur, with élevage in 20–25 per cent new oak. As with many Côte d'Or producers today, the talk is of infusion more than extraction, coaxing rather than forcing the flavour from the grapes. (Or, in cruder terms, let the tea bag rest in the water, don't squeeze it.)

When tasted, the results prompt enthusiastic scribblings, raiding the tasting lexicon to capture the appeal of, for instance, the Nuits-Saint-Georges 1er cru Les Saints-Georges, Nuits' leading *premier cru*, which may one day be elevated to *grand cru*: 'Beautiful wine, full and plump, balanced, lovely texture and finesse, fabulous finish,' according to my tasting note.

Despite the small production, the two white wines are worthy of mention: a *premier cru* La Perrière and a straight Bourgogne Blanc. Both are made from Pinot Gouges, a mutation of Pinot Noir discovered in the family vineyards in the 1930s. The Bourgogne is fresh and lively while the Perrière, whose vines are 80 years old, takes things up a notch. The vineyard is, 'an oven in summer,' according to Antoine and the wine is rich and sumptuous, lavishly flavoured though not cloying.

There has been a gentle lightening of the style at Gouges. The cousins may say it is only a small change but any whose taste buds were

150   CÔTE D'OR

belaboured by the old-style (when not mature) will welcome it. As the Gouges story moves into its second century the domaine is in safe hands.

Try this: Nuits-Saint-Georges 1er cru Les Vaucrains
Gouges owns 1 hectare of this vineyard's 6, a parcel close to the forest with less sun exposure, not an ideal location in the twentieth century but more attractive today for 'it helps to keep tension in warm vintages,' according to Antoine Gouges. Here, the slope is steep and east-facing — and if its origins are a little forbidding the wine is anything but. Beautiful freshness and minerality lead into a wonderfully long finish, with the solid tannins of yore now more supple and attractive.

Domaine Thibault Liger-Belair

40 rue du 18 Décembre, 21700 Nuits Saint Georges
www.thibaultligerbelair.com; tel: +33 3 80 61 51 16

It is difficult to know where to start when trying to elaborate the attention to detail that goes into everything Thibault Liger-Belair does, but his eco-friendly winery on the outskirts of Nuits-Saint-Georges is probably the best place. Built from scratch it saw its first vintage in 2017 and today it gives telling emphasis to his commitment to achieving carbon neutrality. On arrival the first thing one sees is a giant 'solar flower', known as Smartflower, that consists of a fan of solar panels that open and close at dawn and dusk, while the whole apparatus rotates to follow the sun during daylight. It generates enough power to work the winery's air-conditioning.

Every other detail in the new building is similarly impressive, from the glass in the skylights that allows light but not heat into the building, to the sheep's wool insulation in the walls. More specific to the winemaking process are the stainless steel tanks made to his own design. They are roughly square to be more space efficient, with rounded edges and no sharp angles to facilitate circulation of the wine within. And after it is bottled the wire cages normally used to store the bottles before labelling and shipping are abandoned in favour of PVC layers shaped to hold three dozen bottles each, thus obviating the need to wash the bottles after storage.

So far, so sustainable but what of the wines themselves? I wrote previously: 'The results of this approach are impressive: sweet-fruited, satin-textured wines with a whisper of tannin and a streak of minerality to arrest any drift towards lushness.' That assessment still holds good — if anything, the wines are better today, remarkable testaments

THE VILLAGES AND PRODUCERS OF THE CÔTE DE NUITS 151

to Liger-Belair's ability to walk the walk and not just talk the talk. He does both with equal facility, elaborating at length, yet always cogently on every aspect of his winemaking.

'Look for what the grapes can give not for what we can take,' he says, explaining that those grapes are sorted in the courtyard outside the winery and then de-stemmed or not depending on the wine, for instance the Vosne-Romanée Aux Réas is 100 per cent de-stemmed, while the Charmes-Chambertin (below) utilizes 80 per cent whole bunches. Between these, with 20 per cent, is the excellent Bourgogne Rouge Les Grands Chaillots, blended from a hectare each in Nuits-Saint-Georges and Hautes-Côtes de Nuits, and two in Chambolle-Musigny. It is a simple wine in the best sense, with an immediate appeal courtesy of a fresh fruit tingle and delicate intensity.

Liger-Belair's Aligoté Clos des Perrières la Combe – on paper a humble wine – is anything but, being the product of a sophisticated vinification process. The grapes are macerated for one week after foot treading, with the skins held under the juice 'to keep the greenness'. It is then gently pressed and transferred to barrel for 15 months, emerging rich and sumptuous, with Aligoté's signature snap on the palate cushioned by lusher elements. 'It is an Aligoté for food, not aperitif,' says Liger-Belair.

It is for the quality of his top red wines, however, that Liger-Belair is most celebrated and in terms of quality there is little way of distinguishing between those from his own vineyards and those from his small négociant operation, labelled almost identically save for the word 'Successeurs' in small print rather than 'Domaine'. Of the latter, three wines stand out: the Nuits-Saint-Georges 1er cru Les Saints-Georges and the pair of *grands crus*, Clos de Vougeot and Richebourg.

If all goes according to plan the Saints-Georges vineyard will be elevated to *grand cru* in years to come. A process to do so has been in train since 2007 and will likely come to fruition in 2032. 'All the lights are green now,' says Thibault Liger-Belair, who has been at the forefront of the campaign 'to repair an anomaly,' when the vineyard was not given top ranking in the 1930s. There is nobody better qualified than him to champion its cause.

Try this: Charmes-Chambertin grand cru

Charmes is divided from Chambertin itself by the narrow Route des Grands Crus, yet the public's perception of relative quality sees them separated by rather more than that. Which points to the need of always

Thibault Liger-Belair at Les Grands Jours de Bourgogne 2024

asking about burgundy: 'Who made the wine?' In Thibault Liger-Belair's hands this is every inch a *grand cru*. It comes from a plot of 0.3 hectares, with vines dating from 1954, and about 80 per cent whole clusters are used, but with the main stem removed. Aged in 100 per cent new oak from a forest near Chablis, it delivers a cascade of flavour, like a great symphony. Abundant fruit, racy acidity and crisp tannins combine to deliver a complete wine in every sense.

Domaine & Maison Marchand-Tawse

4 & 9 rue Julie Godemet BP76, 21700 Nuits-Saint-Georges
www.marchand-tawse.com; tel: +33 3 80 20 37 32

Not many vignerons have as wide and varied a CV as Pascal Marchand, whose vinous peregrinations have taken him far and wide in the wine world. After stints in charge at Domaine du Comte Armand and Domaine de la Vougeraie he started a small négociant business, making a modest 12,000 bottles across five wines in 2006. He helped to finance this by taking on consultancies around the globe in Australia, California, Canada, Argentina and Chile. Hence, he seemed for a while to slip below the Burgundian radar: 'People thought I had gone away.'

We first met at Vougeraie during harvest 2004. He had been foot-treading a small vat of Musigny, remarking that it deserved kid glove treatment, especially considering the modest quantity. Today, he still turns his deft hand to Musigny (see below), making it in his own name, though now in minuscule quantity. It is the jewel in the Marchand-Tawse crown, though it is a crown of many gems.

It is one of five dozen wines in the Marchand-Tawse portfolio, which divides equally between the Côte de Nuits and the Côte de Beaune. One-third of production comes from the domaine's 8.5 hectares, the remainder from bought-in grapes. Harvest is by hand into 13-kilogram boxes and the grapes are cooled overnight at 5°C, going into the vat at about 8°C, where the temperature is allowed to rise without a pre-fermentation maceration. Marchand largely favours the use of whole bunches, though doesn't have a fixed policy: 'De-stemming is a last-minute decision when we see the grapes on the sorting table.' Sometimes the decision is purely pragmatic, such as in the abundant 2023 vintage, when Marchand had to use fewer stems, simply to fit the harvest into the available volume of vats (it's an observation made by a number of vignerons). During vinification Marchand favours pumping over to punching down, conceding, 'years ago I was known to make big wines,' thrusting his elbows out as he speaks, adding, 'some days we don't even touch a vat.'

This gentle approach is evident in all the wines, notably a trio of Nuits-Saint-Georges: Petite Charmotte and the *premiers crus* Les Perrières and Aux Murgers. The Charmotte's stern core is cloaked by ripe fruit; the Perrières is lean and racy with a pleasant perk of spice; the Murgers takes things a considerable step further in a superbly balanced package where all the elements combine harmoniously into a long finish. At *grand cru* level the Clos Saint-Denis displays a wonderful panoply of contrasting flavours, sweet fruit combining elegantly with savoury notes in a full-textured wine. The Clos de la Roche has all this and then some – a tad more structure and depth giving it greater conviction on the palate.

The Tawse of the name is Canadian, Morey Tawse, whose background is in finance and whose purchase of Domaine Maume in Gevrey-Chambertin some years ago formed the backbone of Marchand-Tawse. The Maume wines were always a bit forbidding, an observation that no longer holds.

154   CÔTE D'OR

Try this: Musigny grand cru
This is a wine of beguiling beauty, whose backstory is worth telling. It comes from a tiny sliver of vineyard that sits on a step at the top of the *premier cru*, Clos de la Perrière, a *monopole* of Domaine Bertagna. Thus it lies completely separate from the main block of the Musigny vineyard, the only clue to its more elevated status being its lofty position. It takes a close examination of a large-scale map to reveal the sliver belongs to Musigny, across the road usually mistaken for the *grand cru*'s boundary. The holding runs to a mere 0.09 hectares, which produces less than a 228-litre *pièce*, necessitating a call to Tonnellerie Remond after pressing to specify the exact volume for a made-to-measure barrel. The wine boasts an abundance of fruit, intense rather than concentrated, in a remarkably delicate framework, and only truly reveals itself on the finish, in the minutes after it has been swallowed. 'It is Amoureuses-ish,' says Marchand.

Domaine de la Vougeraie

7 bis, rue de l'Eglise, 21700 Prémeaux-Prissey

www.domainedelavougeraie.com; tel: +33 3 80 62 48 25

In the same way that Lexus is the upmarket arm of Toyota so Domaine de la Vougeraie is the jewel in the Boisset crown. Founded by siblings Jean-Charles and Nathalie Boisset, Vougeraie celebrated its twenty-fifth birthday in 2024 and it took almost that long for the wines to garner the recognition they deserved. The slightly unusual name didn't help. 'Do you mean Vougeot?' I was frequently asked when telling people of this new kid on the block after my first visit in 2004. Vougeraie has also expanded considerably since the first edition of this book, the domaine now stretching to 52 hectares across 43 appellations. The most significant addition was the 2021 purchase of the 4.5-hectare Clos de la Chapelle in Chassagne-Montrachet from the Domaine Duc de Magenta. It is planted as 3.5 hectares in white and 1 hectare in red. The Magenta white was always serviceable but already the conversion to organic viticulture, certified in 2024, and the introduction of biodynamic practices, are paying dividends with wine that is more expressive on the nose and vibrant on the palate, pure and delicate though not frail.

The expansion in vineyard holdings, together with quicker harvests compared to the more languid pace of yesteryear, has put extra pressure on the grape reception facilities at the winery, necessitating the addition of two extra sorting tables, where previously one served. Somewhat unusually the white grapes, which are picked into 20-kilogram boxes, are also sorted, before going to a pneumatic press, with an overnight settling of the juice in stainless steel. Larger barrels of 350 and 450 litres are used, particularly for the less prestigious appellations, with minimal bâtonnage up to December. Before all this, however, the harvest date is of paramount importance: 'We must pick on the correct day,' says Directrice Générale, Sylvie Poillot.

The red wines utilize varying percentages of whole bunches: in 2023 the Nuits-Saint-Georges Clos de Thorey saw 50 per cent, rising to 80 for the Clos de Vougeot. A cold maceration for four days at 12°C precedes fermentation in wooden vats, with foot *pigeage* once or twice per day. A vertical press, whose 'silky texture' Poillot likes, is then used before the wine is transferred to barrel, about 20 per cent new.

The barrels themselves are worthy of comment. The domaine selects trees from the nearby Cîteaux forest then ages the staves for three years before distributing them amongst five different coopers, each to craft them according to their own favoured methods. When asked what differences may result, given that the raw material used by all five is identical, Poillot simply says, 'They have their own touch.'

The Vougeraie range spans almost the whole of the Côte d'Or, from Gevrey-Chambertin to Chassagne-Montrachet, with an enviable range of *grands crus* in both colours, including: Chevalier-, Bienvenues-Bâtard- and Bâtard-Montrachet, and Musigny, Bonnes Mares and Clos de Vougeot. The latter is a substantial holding of 1.5 hectares in two parcels, a hectare near the top, across the road from Le Clos Blanc (below), and the remainder at the bottom, running down to the D974 road. The two are vinified separately and then blended and allowed to 'marry' in stainless steel for two months before bottling. It is a fine, elegant wine with the weight and depth of flavour one expects from a *grand cru*.

Try this: Vougeot 1er cru Le Clos Blanc de Vougeot monopole
This vineyard extends to about 2.25 hectares in a roughly triangular plot, marked by a modern gateway at its upper end, close to the entrance to the Château du Clos de Vougeot. This is Vougeraie's flagship white wine, though the 'Blanc' in the name does not refer to the wine but to the pale colour of the vineyard's soil, first planted with Chardonnay for sacramental purposes by the Cistercians about 900 years ago. Today, 4

per cent Pinot Gris and 1 per cent Pinot Blanc underscore the wine's opulence and satin texture, leavened by a leaner flavour dimension courtesy of a 'tannic' dryness and a saline snap on the finish.

5

THE VILLAGES AND PRODUCERS OF THE CÔTE DE BEAUNE

The Côte de Beaune is expansive by comparison with the more compact Côte de Nuits, a contrast signalled by the broad-bellied Hill of Corton that reclines on the horizon as you drive south from Corgoloin and Comblanchien, leaving the stone and dust of the quarries behind. The hill itself could act as a metaphor for the Côte d'Or, its glories and its challenges: some of its wines, especially the best Corton-Charlemagnes, are ranked amongst Burgundy's finest, while some of the red Cortons plod on the palate in a way that questions their *grand cru* status, though such disappointments are fewer now than heretofore.

The physical contrast with the Côte de Nuits is immediately apparent. The Côte de Beaune is expansive, sprawling even, compared to the linear succession of communes in the Côte de Nuits. There is a second, inner, line of villages set well back from the spine of the D974: Pernand-Vergelesses, Savigny-lès-Beaune, Monthélie, Auxey-Duresses, Saint-Romain and Saint-Aubin. Where the Côte de Nuits is seldom more than a kilometre wide, except at Gevrey-Chambertin, the Côte de Beaune frequently spans two kilometres and stretches to five or six between Meursault and Saint-Romain, or between Puligny-Montrachet and Saint-Aubin.

The hill of Corton is a plump lozenge capped by a tuft of forest that looks like a toupee. Three villages, Ladoix-Serrigny, Aloxe-Corton and Pernand-Vergelesses, surround the hill like moons around a planet,

158 CÔTE D'OR

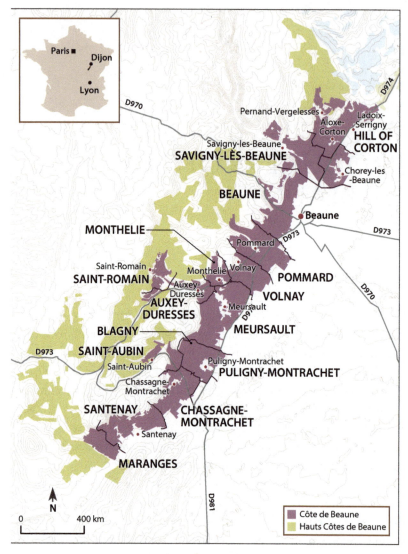

Map 7: The Côte de Beaune

each claiming a share of its *grands crus* vineyards within their commune boundaries, though only Aloxe claims its name. South of the hill and past Savigny-lès-Beaune and Chorey-lès-Beaune, the A6 motorway slices across the grain of the côte, in much the same way that the lesser N6 cuts through further south, separating Chassagne-Montrachet from its most prestigious vineyards.

Crossing the motorway to Beaune, the vineyards retreat up the hillsides to make way for light industrial zones and bland suburbia.

It's only when south of the vinous capital that you come to a similar string of villages to the Côte de Nuits, each with a clear identity in consumers' minds: Pommard, Volnay, Meursault, Puligny-Montrachet and Chassagne-Montrachet. The first two produce exclusively red wines and the latter trio predominantly white wines – the greatest whites of Burgundy, many would say the world. If the Côte de Beaune has a centre of gravity it is the little intersection at the north-east corner of Montrachet, where it is worth standing for a few minutes at the Laguiche entrance to the vineyard to survey the expanse of fabled vineyard stretching in every direction.

Nothing more clearly illustrates the difference from the Côte de Nuits than the two dog-legs that swing away from the main drag, the first narrowing to a connective spindle of vineyard before broadening at Saint-Romain, the second sweeping away from Puligny and Chassagne to Saint-Aubin. After that the wine changes back to predominantly red as the Côte d'Or swings to a finish through Santenay and on to Maranges.

THE HILL OF CORTON: LADOIX-SERRIGNY, ALOXE-CORTON AND PERNAND-VERGELESSES

Whenever an example is needed to illustrate the inconsistencies of the Côte d'Or's vineyard classification Clos de Vougeot is lined up for a ritual beating, pointing out the nonsense of such a large vineyard being granted a blanket right to *grand cru* glory. This must please the burghers of Ladoix-Serrigny, Aloxe-Corton and Pernand-Vergelesses no end – all three of which have vineyards boasting the Corton name in some shape or form. Red Corton is the Côte de Beaune's equivalent to Vougeot, its twin in many respects, not all of them complimentary. Perhaps because Corton is not a *clos* and therefore is not as easily identified as Vougeot it escapes the latter's fate as pantomime villain. Granted, in the best hands and from the best sites such as Les Bressandes or Clos du Roi, Corton can be exciting and satisfying but it is too often on the limp side, without the intensity and nuances of flavour that a *grand cru* should possess.

Ladoix-Serrigny is the 'irregular verb' of the Côte d'Or's hyphenated nomenclature, being the conjunction of two village names rather than

160 CÔTE D'OR

village and prized vineyard. Physically it is strung out, without the cluster of dwellings surrounded by vineyards seen in most villages. Aloxe-Corton is different, off the main road and tight-packed at the base of the hill of Corton where its narrow streets wriggle close to *grand cru* vineyards. Finally, Pernand-Vergelesses is tucked away, around the corner as it were, where its hilly streets look up to vine-clad slopes on all sides. It may sound trite, but for non-French speakers 'Pernand-Vergelesses' is quite the mouthful, a circumstance that doesn't help in gaining what might be termed verbal visibility. In wine terms it is also hard to define, for some of the vineyards face directly east, looking across the small valley to others that face directly west. Thus it is difficult to give a catch-all description of the commune's style, however lazy that might be. In short, the triumvirate of Ladoix-Serrigny, Aloxe-Corton and Pernand-Vergelesses lie a little below the consumer radar and it is only savvy Burgundy lovers who know that astute buying can yield both quality and value.

Producers

Domaine Bonneau du Martray

2 rue de Frétille, 21420 Pernand-Vergelesses

www.bonneaudumartray.com; tel: +33 3 80 21 50 64

Bonneau du Martray dates from post-Revolutionary times and was owned by the le Bault de la Morinière family until early 2017, when a majority holding was bought by Stanley Kroenke from America, whose other wine interests include Screaming Eagle in Napa Valley. The first edition of this book was published shortly afterwards, meaning that, apart from noting the purchase, little could be said about changes in practices or style following the acquisition. In the years since much has indeed changed.

Most notably, about 3 hectares of vineyard have been leased to Domaine de la Romanée-Conti, leaving 8 in the charge of the same winemaking team, now led by manager Thibault Jacquet, who worked previously for Laurent-Perrier in Asia. 'Having the same size team for a smaller domaine allows us to be more precise,' says Jacquet, going on to note that, while the vineyard holdings are not minutely parcellated, the winery facilities are scattered across ten separate locations in Pernand-Vergelesses: 'The team spend their lives moving things.' Moves are in hand at time of writing to partial-ly rectify the situation, renovating the premises at rue de Frétille to double the space, though all work must be sensitively executed, as the buildings are protected.

Out of such complication comes simplicity – in the shape of a pair of wines, Corton-Charlemagne in white and Corton in red. Truth be told, the domaine's reputation rests almost exclusively on the Corton-Charlemagne, from whose shadow the Corton has seldom escaped. After harvesting the Chardonnay grapes into small 10-kilogram boxes, vinification starts with a gentle crushing and a slow pressing followed by two days' *debourbage* at ambient temperature. Fermentation starts in stainless steel before transfer to 600-litre barrels and some amphorae, 'not because they are awesome, it's simply part of the process.' A modest 20 per cent new oak is used and there is no bâtonnage: 'We don't want to be violent.'

Ten separate blocks make up the Corton-Charlemagne whole and these are vinified separately, even to the extent of using different barrels for different sections of the vineyard. François Frères is the favoured cooper for the richer, downslope plots, while Remond barrels are used for the more mineral upslope plots 'to give more shoulders'. Thus 10 separate wines are made, before blending in the June after harvest. Any barrels deemed to be sub-standard are sold off to négociants.

The Corton-Charlemagne may yet have a rival from within: 'We are looking for new challenges, other opportunities, beyond the Hill of Corton,' says Jacquet, explaining that a few rows of Bâtard-Montrachet, amounting to 0.25 hectares, were bought at the end of 2022, and he is on the lookout for other parcels of similar pedigree.

There is also a challenge closer to home: if the team can bring the perfectly serviceable Corton up to the quality of the Corton-Charlemagne then that will be a real achievement and the change of ownership will be assayed a triumph.

Try this: Corton-Charlemagne grand cru

The expansive band of vineyard that wraps around the Hill of Corton, forming a rough horseshoe shape, is divided between almost 100 owners, nearly five dozen of whom lay claim to less than 1 hectare. Not so Bonneau du Martray, the second largest owner, whose holdings straddle the boundary between Aloxe-Corton and Pernand-Vergelesses, orientated south-west to west, thus catching the evening sun. Since the change of ownership, the discreet style of old has yielded a little to a more expansive, opulent palate, while not sacrificing the purity and precision

162 CÔTE D'OR

of flavour. Refinement and reserve are complemented by glossy fruit, echoing long into the finish.

Note: This wine is named simply as 'Charlemagne' from the 2021 vintage.

Domaine Rapet Père et Fils

3 Impasse de la Mairie, 21420 Pernand-Vergelesses
www.domaine-rapet.com; tel: +33 3 80 21 59 94

Domaine Rapet is tucked away discreetly down a narrow road, well hidden from passing traffic, a location that mirrors the quiet presence of the wines in the market. This is not a headline, or headline-seeking, domaine, yet its roster of vineyard holdings suggests it should be better known. There are 3 hectares of Corton-Charlemagne spread across seven plots, mainly west-facing, and over a hectare of Corton, which includes half a hectare of the Les Pougets *climat*; add in a slew of *premiers crus* in Pernand and Beaune, together with a range of communal holdings, and you get a significant domaine of some 25 hectares.

It traces its origins back to 1765, though the domaine as currently constituted is the work of Vincent Rapet, who started here in 1985 and whose son, Robin, has now joined him. The winery dates from 2003, allowing it to accommodate the 350-litre barrels used for the white wines (currently fashionable larger barrels can present problems in cramped, though picturesque, old cellars). New oak is used sparingly: about 10 per cent for *village* wines, 20 per cent for *premiers crus*, and 33 per cent for the Corton-Charlemagne.

This latter wine is the standard-bearer and, after pressing and a 24-hour cool settling, is fermented principally in barrel, with one 12-hectolitre foudre, one 8-hectolitre terracotta amphora and one 6-hectolitre concrete egg utilized for the 2023 vintage. The result is a lush, plump, suave-textured wine with some mild spice to hold the fuller flavours. 'It is a shy wine when it is young,' according to Robin Rapet, 'needing at least five years' before it shows well.

Its sibling red, the Corton Les Pougets, comes from a south-facing vineyard on the mid-slope at about 250 metres elevation. After sorting, one-third whole clusters are used in the fermentation, which lasts about two weeks, with regular *pigeage*, and ageing takes place in barrels, of which 20 per cent are new. The result is a dark, firm-structured wine with ample depth of flavour that is well capable of ageing – approaching its thirtieth birthday, the 1995 was showing a bit of sinew as the fruit faded and the tannin asserted itself but there was still good intensity and reasonable length.

The Rapet range comprises two-dozen appellations, and amongst their number is included a fresh Chorey-lès-Beaune with a lovely cherry rasp of fruit, but it is the *grands crus* that carry and burnish the domaine's reputation. It is also worth noting the excellent website, which is a great resource for further exploration of this wide range of wines.

Try this: Bourgogne Aligoté

Robin Rapet notes that Pernand-Vergelesses is, 'an historic village of Aligoté', going on to explain, 'it is complicated to get good maturity in Aligoté, the window is short.' The grapes come from the regional vineyard La Grande Corvée de Bully on the north face of the Hill of Corton at an altitude of 350 metres. Such unprepossessing origins deliver a crisp, perky wine, vivid and challenging, calling the senses to sit up and take note. It is fresh, with a pleasant sharpness that calls to mind grapefruit. As a food match the domaine's website avers: 'It will coax any Burgundy snail out of the shell to its place at the table.'

SAVIGNY-LÈS-BEAUNE AND CHOREY-LÈS-BEAUNE

Savigny and Chorey face one another across the D974, which forms the spine of the Côte d'Or, with the Hill of Corton rising to the north and Beaune itself to the south. Of the two, Savigny is the more rewarding to visit and is but a short drive from Beaune, though its situation close to the wooded hills makes it feel further away and considerably more rustic. Along with other villages such as Marsannay and Maranges its star has been in the ascendant in recent years, the useful though unexciting red wines of yore now replaced by more vividly flavoured and memorable ones. Savigny enjoys greater visibility than its neighbour Pernand-Vergelesses, probably thanks to the 'Beaune' in its name, and the village itself straddles the modest river Rhoin, tight packed against its banks by the surrounding hills.

The Chorey-lès-Beaune commune slots like a dovetail between Savigny and Aloxe-Corton, and the village lies on the flat land to the east of the main road. Without any pretensions to *grand* or *premier cru* prestige, Chorey continues to do what it has done well for decades

164　CÔTE D'OR

– produce commendably fresh red wines whose attraction divides equally between the contents of the bottle and the modest hit on the contents of your wallet. For savvy consumers Chorey continues to deliver some of the best value in the côte.

Producers

Domaine Chandon de Briailles

1 rue Soeur Goby, 21420 Savigny-les-Beaune
www.chandondebriailles.com; tel: +33 3 80 21 52 31

Chandon de Briailles has been family-owned since 1834 and today is in the charge of seventh-generation brother and sister François de Nicolay and Claude de Nicolay-Jousset who, since taking the reins in 2001, have done much to improve the quality and reputation. Biodynamic practices were introduced in 2005 and the domaine has been certified since 2011. One unique treatment to combat oidium, while reducing the use of sulphur, is to spray with an 80:20 mix of water and low fat milk. Four hundred litres of *petit lait*, supplied by a cheesemaker, are used and, having described the process, Claude Jousset concludes: 'I would love to find the right way to avoid using copper.'

The challenges and current benefits of climate change dominate Jousset's conversation: 'We have to adapt, we need to move further on how the vines are treated … All the changes we have made are outside, the cellar is much the same … Global warming has helped with better ripeness in cold vineyards, we are still in a beneficial period … The new climate helps the fruit, especially the tannins, we don't see austere wines now … We need more pickers at harvest time because the picking window is shorter … Dryness is a real issue here … How are we going to manage without water?' One telling fact neatly illustrates the significant change witnessed by all vignerons in recent years: at Chandon de Briailles 2020 was the first vintage that not only started in August but even more unusually also finished in August. Notwithstanding the challenges, the Chandon de Briailles wines have gained assurance and conviction over the past decade.

Vineyard holdings stretch to about 14 hectares, principally around the hill of Corton, with a small outpost in Volnay Caillerets. Red wines account for three-quarters of production. Whole bunches are generally used because they give, 'elegant and refined wines,' but the decision, 'depends on the ripeness of the fruit and the stems, it is a different decision

THE VILLAGES AND PRODUCERS OF THE CÔTE DE BEAUNE 165

every year.' The grapes are gently cooled if necessary and fermentation starts in open-topped vats after a few days, with pumping over favoured at first, then punching down. A 14- to 18-month élevage in barrel follows before bottling without fining or filtration.

These methods delivered impressive results in the difficult 2021 vintage, when cool damp conditions for much of the summer challenged the vignerons after the previous trio of hot vintages. The Savigny-lès-Beaune 1er cru Les Lavières, made with 50 per cent whole bunches and a yield of only 12 hectolitres per hectare, gives the lie to the commonly held perception that 2021 wines are sub-standard – it was quantity that was badly affected, not quality, which held up well. One sip of this exuberant Savigny makes the point convincingly.

Concluding my profile of Chandon de Briailles in 2017 I wrote: 'The Corton-Bressandes ... is testimony to what has been achieved at Chandon de Briailles since the turn of the century, yet one senses that the best days are still to come.' Those days have now arrived.

Try this: Corton-Bressandes grand cru

If ever the finest *climats* on the Hill of Corton are separated out and granted appellations in their own right, Bressandes will be at or near the top of the list. It is perfectly sited, gently inclined on the mid-slope and faces east with a slight turn south. Chandon de Briailles owns four parcels, amounting to 1.5 hectares. Made with 100 per cent whole bunches, the wine majors on ample sweet fruit with tannin and acid whispering in the background. It relies on balance and harmony rather than concentration and weight to make a lasting impression.

Domaine Tollot-Beaut

Rue Alexandre Tollot, 21200 Chorey-lès-Beaune

Tel: +33 3 80 22 16 54

Speaking of the 2023 vintage, Nathalie Tollot neatly encapsulates the vicissitudes of the vigneron's life: 'In late August we were anxious about lack of ripening. Then there was a heatwave in early September and one week later we were anxious about over-ripening. When you picked made a big difference.' To avoid the worst of the heat they started to pick the Chorey-lès-Beaune at 6.30 in the mornings.

Tollot-Beaut employs a team of 60 to 70 pickers at harvest, who need to work faster than in previous years, 'because the harvest

Harvest 2023 at Domaine Tollot-Beaut. Pickers in the vineyards at 6.30 a.m.

window is narrower.' Then, as is the domaine's practice, the grapes are 100 per cent de-stemmed – gently, so that whole berries with uncrushed pips go into the stainless steel and concrete fermentation tanks. Some punching down follows over the next fortnight and the wines rest for a few days in tank after fermentation, before élevage of about 16 months in standard 228-litre oak barrels, 25–50 per cent new, depending on appellation.

The results seldom, if ever, disappoint. Utterly reliable is the Savigny-lès-Beaune 1er cru Les Lavières, a large, south-facing vineyard in which Tollot-Beaut hold a generous two hectares. Some of the vines date from 1943, though a large portion was replanted in 2017. Nathalie Tollot makes the point that they don't want to stress the vigorous young vines because, 'we want to keep them for 60 or 70 years'. They are planted on a thin layer of clay, some 30–40 centimetres deep, and the wine is splendidly intense with ample fruit and a mild tannic grip in a smooth-textured package.

Unless enunciated correctly, saying that Tollot-Beaut is probably the most dependable name in all the Côte d'Or can sound like damning with faint praise; it is hard to avoid an unintended pejorative tone. Yet that is precisely this domaine's calling card – few rival it as a longstanding benchmark for value and consistency. Tollot-Beaut delivers again and again without recourse to attention-grabbing practices or glitzy marketing spiels. Indeed, the cudgel-shaped bottles used today, that replaced the standard Burgundy shape some years ago, together with their utilitarian labels, seem to go out of their way not to charm. Rest assured, however, that shape and label belie the charm and grace of the contents. These 'books' are not to be judged by their covers.

Try this: Beaune 1er cru Les Grèves

Beaune's over-population with *premiers crus* detracts somewhat from the standing of those fully deserving of the designation. One such is Les Grèves, which is noted for producing the most 'serious' Beaune of all. It is a large vineyard of 31 hectares, with more than two dozen owners, and the Tollot-Beaut holding of 0.59 hectare is one of the domaine's oldest parcels. 'It is always more gentle with aeration,' says Nathalie Tollot, pouring from a bottle 'open since yesterday, so it's better'. And it is, less forbidding, with the dark fruit and severe structure lightened by contact with air. One for the cellar.

BEAUNE

In general, Beaune wines are more pleasant than profound and a sprawling surfeit of *premiers crus*, far from elevating its status rather diminishes it, consigning it to a 'safe and sound' reputation in consumers' minds. The vineyards lie to the west, on a bank of gently sloping hillside that rewards exploration on foot. The town of Beaune is Burgundy's bullseye: Dijon may be the administrative capital of the region but Beaune is the wine capital, a circumstance that is impossible for even the least wine-minded visitor to miss. Beaune is built on wine, literally: there is barely a cobbled street in the town centre that doesn't sit above a vaulted stone cellar.

There could hardly be a greater contrast between the historic centre and the outskirts, where medium-rise apartment blocks stand beside giant supermarkets. The suburbs are featureless, unlike the attractive centre, though even here opinion is divided. Some see it as little more than a twee tourist trap while others revel in its energy, most obviously on display on a Saturday morning, market day. Tourists abound during the summer but there are plenty of locals shopping too, including well-known winemakers. It's trite, and untrue, to dismiss this as little more than a tableau to dupe tourists.

Thanks to its scale Beaune sits apart from its eponymous vineyards, unlike the smaller villages, where it is often not possible to say

168　CÔTE D'OR

where vineyard ends and village begins, or vice versa. It has the feeling of a grand metropolis when compared to the likes of Chassagne or Chambolle and the other tiny villages whose reputations exceed it in purely vinous terms. Many of the large négociants have their head-quarters in Beaune: Drouhin, Jadot, Latour and others, known trad-itionally for their skill as blenders rather than makers of wine. Blending can eliminate character and when a house style needs to be maintained individual vineyard identity is further eviscerated. Thus the received reputation of Beaune wines for delivering moderate satisfaction prob-ably derives from the fact that so much of the vineyard land has been in négociant ownership for decades and longer. Quality was sound, not inspiring; boats weren't rocked or perceptions challenged; the house style marched on. Beaune wines seldom set the pulse racing; they were reliable fillers on a wine list, destined to sing in the chorus, never to take centre stage.

Blame for the lack of lustre cannot all be laid at the négociants' doors for, as mentioned above, Beaune has an unfeasibly large proportion of *premier cru* vineyards, over 330 hectares being so designated, compared to about 140 of *village*. This ranks as a serious aberration and skews the reputation, thanks to nominally superior vineyards delivering little more than *village* quality. With the *premier cru* currency thus debased one must exercise more than the normal level of caution when buying, but the good news is that the négociants' reputation for blandness is increasingly historical.

It is difficult to generalize about the wines in the way that you might about Gevrey or Meursault, but their first duty should be to charm the palate by way of vivid fruit and juicy tingle. Above all Beaune should deliver fruit in bountiful measure – and from *premier cru*, fruit backed by more substance and structure. A good Beaune is charm personified; profundity can be found elsewhere.

Producers

Domaine Bouchard Père & Fils

15 rue du Château, 21200 Beaune

www.bouchard-pereetfils.com; tel: +33 3 80 24 80 45

Bouchard traces its history back to 1731, when Michel Bouchard, who traded in fabric, switched occupations to embark on a career in wine. He was succeeded by his son, Joseph, who bought the family's first vine-yards in Volnay in 1775. Inevitably, Bouchard benefited from the sale of church and aristocratic lands after the Revolution and continued to acquire vineyards over the succeeding generations. By the mid-twentieth century the firm was riding high, but decline followed, leading to the sale of the business to Joseph Henriot from Champagne in 1995. It is fair to say that, notwithstanding such an ancient lineage, more change and transformation has been packed into the 30 years since than occurred in the previous centuries.

The Henriot acquisition helped to revitalize Bouchard, though the change did not occur in time to prevent the 1996 Montrachet from being corrupted by premox. Nevertheless, the 2005 construction of a new winery in Savigny-lès-Beaune boded well for the future and an all-round improvement in quality was noted. Then, in 2022, it was announced that the Artémis group, owners of Clos de Tart (see p. 102) and Domaine d'Eugénie amongst others, had merged with Bouchard, keeping Henriot as a minority partner. That arrangement lasted about a year, before Champagne Henriot and William Fèvre Chablis (owned by Bouchard) were both sold, leaving Artémis in sole charge of a trimmed-down Bouchard, now reborn as a domaine, rather than a négociant.

Since then change at Bouchard has been swift, including the installation of air conditioning in the historic Beaune cellars, which date from 1846 and whose one kilometre of tunnels and galleries houses three million bottles. Despite being at a depth of 10 metres it was noted that the regular temperature of 11–13°C climbed to 17°C in recent hot summers. In addition, Bouchard has narrowed its focus and is concentrating solely on the Côte de Beaune, the Côte de Nuits vineyards being transferred to Eugenie. Cellar door sales have also ceased.

At time of writing Bouchard is a work in progress – delivering definitive judgement on a producer whose practices are in a state of flux would be like commenting on a painting as the artist is working on it – though it can be stated emphatically that change is in the air, and it appears to be for the good. Even stripped of its Côte de Nuits treasures such as Chambertin and Bonnes Mares, Bouchard's vineyard holdings are impressive, especially in white: not quite a hectare of Montrachet, well over two of Chevalier-Montrachet, four of Corton-Charlemagne and, in red, over three of Corton – not forgetting the favoured child that is the Beaune Grèves *premier cru* Vigne de l'Enfant Jésus. This trimming has reduced the vineyard holdings from 130 to 105 hectares

170   CÔTE D'OR

Hopefully the move to domaine status will rid the wines of the cloak of homogeneity that used to be their calling card. It needs to happen and, given the recent pace of change, it will probably happen quickly. It cannot be too quickly.

Try this: Beaune du Château 1er cru

This wine has been made since 1907 and if it was a political party it would include members from every hue and spectrum of society, for it is an amalgam of 17 different *premier cru* plots that reads like a roll-call of Beaune vineyards, from tiny Les Seurey (1.23 hectares) to grand Les Cent Vignes (23.5 hectares). Each lot is vinified separately and then blended, before spending 14 months in barrel, 30 per cent new. The result is a wine that majors on cherry fruit, mouth-watering and zippy, rather than profound. A sibling white is made from a more manageable five plots.

Domaine Chanson Père & Fils

10 rue du Collège/rue Paul Chanson, 21200 Beaune
www.domaine-chanson.com; tel. +33 3 80 25 97 97

It is a quarter of a century since Champagne Bollinger bought Chanson and began a thorough shake-up of this historic but humdrum name, whose wines at the time all tasted much the same, regardless of the label. Much hoopla surrounded the acquisition and a noticeable uptick in quality followed but in recent years there was a sense that this had stalled. It has now kicked in again and a new dynamism was signalled with the appointment of Vincent Avenel as managing director five years ago, and the purchase of 50 hectares of Côte Chalonnaise vineyards in Mercurey in 2023.

'It is important to be humble but also ambitious,' says Avenel as he outlines the company's recent change of strategy. Previously, about 25 per cent of production came from the domaine's own vineyards with the remainder made up of purchased grapes or must. After the Mercurey acquisition the domaine to négociant ratio is now about 50:50. This significant shift was partly forced on them by the succession of small vintages that ran, roughly speaking, between the abundant and high quality 2009 and the similarly abundant but more workmanlike 2017. In that period it became increasingly difficult to secure good quality material for the négociant wines, and the realization that the situation was unlikely to improve (witness the low quantity, good quality 2021 vintage) prompted the change.

THE VILLAGES AND PRODUCERS OF THE CÔTE DE BEAUNE 171

Avenel comes with the right credentials, having previously worked at Domaine Faiveley when it was undertaking a similar transition. Currently the wines still carry some of the safe and sound character that marked them in recent years, though there is one that points the way to what might be achieved in the near future: the Chassagne-Montrachet 1er cru Les Chenevottes, an astute purchase in 2006. With a 1.9-hectare holding, Chanson is the largest owner, although only half is currently in production, the other recently replanted half being declassified to Bourgogne Blanc. There is nothing bland or safe about this Chenevottes, it is crisp and intense, creamy and tingling in equal measure, and departs with a pleasant saline whiff on the finish. Bring all the wines up to this level of quality in their respective categories and Avenal will have done handsomely.

Though the Chanson wines are made in modern premises near Savigny-lès-Beaune, the headquarters and ageing cellars remain in the centre of Beaune, in the largest of the five *bastions* that formed part of the town's defences. Originally named La Tour des Filles it was renamed Bastion de l'Oratoire and thanks to its seven-metre-thick walls it provides a perfect above-ground 'cellar' over four levels. The core vineyard holdings remain the 43 hectares in the Côte de Beaune with a heavy preponderance of Beaune *premiers crus*. These are currently undergoing a change in style occasioned by a reduction in the percentage of whole bunches used in the vinification. Previously, it used to be 100 per cent but today that figure is more likely to be 50. Avenal explains: 'Ripe stems give more spice and floral complexity, structure and more power. If too many whole bunches are used it might give a green structure, it might be too strong.'

If Chanson can achieve the same transition as that effected at Domaine Faiveley over the last 20 years then, allowing for a similar timescale, they will be in a good place by mid-century. The company marks its two-hundred-and-seventy-fifth anniversary in 2025. Celebrations are in order, though if current progress continues they will be dwarfed by the three-hundredth in 2050.

Try this: Beaune 1er cru Clos des Fèves monopole

Clos des Fèves is a 4-hectare *monopole* within the Les Fèves vineyard, the only other owner of which is Château de Meursault. It is located at the northern end of the commune, lying at 250 metres altitude, downslope of the better-known Les Bressandes and above Les Cents Vignes. An

172   CÔTE D'OR

exotic nose presages a structured, robust palate replete with strident tannins. The fruit is dark and brooding and calls for a decade in the cellar to endow grace and harmony. Only then will it charm as well as satisfy.

Maison Joseph Drouhin

7 rue d'Enfer, 21200 Beaune

www.drouhin.com; tel: +33 3 80 24 68 88

There is a marked contrast between the Drouhin cellars in the centre of Beaune, parts of which are built upon Roman walls that date from 380, and the utilitarian premises that house the modern winery on the outskirts of the town, with a Second World War Sherman tank standing sentinel nearby. Clichéd as it is, 'atmospheric' is the appropriate descriptor for the underground labyrinth that covers a whole hectare; a ball of twine would not go amiss were it to be explored solo.

The business was founded by the eponymous Joseph in 1880 and is now run by the fourth generation, siblings Philippe, Véronique, Laurent and Frédéric. Under their stewardship the wines have gained distinction and identity, avoiding 'négociants' malaise' – the affliction that sees all the wines from a big négociant tasting much the same, carrying a bland overlay that partially obscures their vineyard and vintage origins. Such a criticism might have applied at the turn of this century but not anymore.

Véronique Drouhin, who oversees winemaking, both in Beaune and at their American outpost, Domaine Drouhin Oregon, is mindful of that trap: 'Sense of place is very important for us ... I think of Meursault as broad and Puligny as tall ... During élevage we taste all the wines every Monday ... they should show typicity ... If a wine is far from what we want, we sell or declassify it.' As an aside, Véronique could justifiably claim that her future role was written in the stars, for she was born on the day of the Hospices de Beaune wine auction and a week later was treated to a 'proper christening', when a drop of Musigny 1856 was placed on her tongue.

Today, her wealth of experience gives her a long-term perspective, so she is not blind to the cyclical nature of the wine business: 'I saw my father struggling, stressed every day to make sales ... It changed in 1985 ... We have been unbelievably lucky with good vintages and crazy demand since then.' Drouhin is also sanguine about the difficulties posed by the 2024 vintage, saying, 'It is a vintage like my father had most years.' In addition, she is cognizant of the influence of climate

THE VILLAGES AND PRODUCERS OF THE CÔTE DE BEAUNE   173

change, although she points out: 'The Drouhin style has not changed. Our philosophy is for elegant wines. We have gained structure without compromising finesse.'

There is finesse aplenty in the wines, illustrated to good effect by the Pommard Chanlins-Bas, a relative newcomer to the Drouhin stable, the plot on the Volnay side of the commune having been bought in 2014. It is an elegant wine with a rounded texture and refined tannins and is not in any way rustic or lumpen. A white equivalent is the Puligny-Montrachet 1er cru Clos de la Garenne, a small vineyard that lies below the hamlet of Blagny. It is intense but not heavy, svelte textured with a savoury, saline finish.

More recently, the Drouhin stable has expanded with acquisitions in Saint-Romain and Saint-Véran, adding 20 hectares and meaning that Drouhin's holdings now stretch all the way from Chablis to the Mâconnais. For now, the future of Drouhin looks secure, though whether the current quartet of siblings will be succeeded by any of the seven members of the next generation is not yet certain.

Try this: Beaune 1er cru Clos des Mouches Blanc

'It is the cherished child of the family, combining power and elegance,' says Véronique Drouhin, with no word of exaggeration, when speaking of Clos des Mouches. It is worth adding a caveat, however: drink this wine too young and you will be left wondering what all the fuss is about; catch it when properly mature and you will think you are in *grand cru* territory. Drouhin owns over half of this 25-hectare vineyard, which lies on the southern boundary of the commune, bordering Pommard. There is also a red version but it is the white that gains the most enthusiastic plaudits. It delivers a cornucopia of citrus and tropical fruit flavours that crosses the palate like a rolling wave, with a saline, mineral twist to add complexity and enduring attraction.

Maison Louis Jadot

62 route de Savigny, 21203 Beaune

www.louisjadot.com; tel: +33 3 80 22 10 57

'At harvest, the window of picking is shorter and shorter. Since 2018 we have used three teams totalling 180 pickers, whereas it used to be two, totalling 120. The question is not when you start but how soon you finish.' Frédéric Barnier has overseen winemaking at Jadot since 2013 and his harvest observations are echoed at other producers, if not on

174 CÔTE D'OR

the same scale. Talking of recent growing seasons: 'The weather is more complex now with greater inconsistencies; 2023 was more challenging than 2022, but it was nothing compared to 2024.' It is still too early to pass definitive judgement on the 2024 vintage but when visiting dozens of domaines that summer the vignerons were almost unanimous in observing – wearily – that it was one of the most challenging seasons in memory.

Jadot was founded in 1859 by Louis Henry Denis Jadot and was owned by his descendants until 1985, when it was sold to the Koch family from the United States, owners of Kobrand, Jadot's US importer. A third family has also been closely associated with Jadot since 1954, when André Gagey joined as deputy to Louis Auguste Jadot, taking over from him on Jadot's sudden death in 1962. André was succeeded by his son Pierre-Henri, and his son, Thibault, joined the company in 2014 and is now general manager, while the man in overall charge today is Thomas Seiter who joined Jadot from Bouchard Père & Fils in 2022.

Jadot has control of 130 hectares of vineyard in the Côte d'Or, some owned directly and others owned by entities such Domaine André Gagey and by members of the Jadot family through Les Héritiers de Louis Jadot. The jewel in that latter crown is the Beaune *premier cru* Clos des Ursules, a *monopole* within the Vignes Franches *lieu-dit* which was bought by Louis Jadot in 1826. A bicentenary beckons. The vineyards stretch from Marsannay to Santenay, with a concentrated cluster, amounting to 25 hectares, in Beaune. From these last, twenty-two separate wines are made, nearly all of which are *premiers crus*. Some, such as Bressandes and Grèves, are well known, others like Chouacheux and Pertuisots, less so.

Centre of operations is the modern winery in the Beaune suburbs and, as noted above, the increased pace at which the grapes arrive presents a new challenge, met in part by the recent construction of a large cool room, 'a waiting room,' as Barnier quips. For the white wines, the hand harvested grapes will be cooled if necessary before going straight to a pneumatic press and then to barrel, 25 per cent new from Jadot's own *tonnellerie*, Cadus, with no settling of the juice. Malo can be blocked if necessary to 'maintain a certain freshness,' during the 15- to 18-month élevage. Then into stainless steel tank for one to two months before bottling.

For the reds, the grapes are mostly de-stemmed, about 90 per cent, after hand sorting, and then fermented in open-top oak and stainless steel vats. There are two punch-downs per day with no pumping over. From *village* to *grand cru* the élevage is almost the same and sees increased use of 450-litre barrels and large foudres of various capacities to reduce the oak influence. 'We are changing the way we age our wines to make sure the wines are not changing,' says Barnier.

Production divides about 65:35 in favour of red wines, spread across a mind-numbing range – 212 wines are listed on the website – with exports accounting for 85 per cent of production, to 120 countries, led by the USA, UK and Japan. With some exceptions, notably the Chevalier-Montrachet Les Demoiselles and the Clos de Vougeot (below), the house style is satisfying rather than exciting, though it has moved a little in the latter direction in recent years.

Try this: Clos de Vougeot grand cru

Too often in the past I have found myself expecting more outright class from Jadot's *grands crus*: 'sound but not spectacular,' was the gist of many tasting notes. If today's Clos de Vougeot is anything to go by those days are receding. The Jadot holding, at over 2 hectares, is one of the largest and is in one plot – making it easy to spot on a map of this pixelated vineyard. The vine rows are 400 metres long and although they lie in the lower, less favoured section of the vineyard they produce a wine of commendable intensity with well-rounded tannins and an earthy character that avoids overt rusticity.

Maison Louis Latour

18 rue des Tonneliers, 21204 Beaune
www.louislatour.com; tel: +33 3 80 24 81 10

Though the address above places Latour in the centre of Beaune, this is for administrative purposes only; the company's historic heart lies a few kilometres to the north on the Hill of Corton. There, surrounded by vineyards, is found the Cuvérie Corton Grancey, tucked into a declivity that was once an old quarry. It is an imposing block that matches the squat form of the hill and was built between 1832 and 1834. Latour bought Grancey in 1891 and today it functions as the winery for all the Latour domaine wines. It is gravity fed over five levels and even the doors utilize the earth's pull, closing with a mighty slam, thanks to a weight-and-pulley apparatus.

The early death in 2022 of Louis-Fabrice Latour, seventh generation to take charge of this family firm, was widely mourned in the

176   CÔTE D'OR

wine world. As president of the Bureau Interprofessionnel des Vins de Bourgogne from 2015 to 2021 he was for many the affable face of Burgundy: tie loosened and with broad smile always threatening to break out. The family succession has paused for now, though his daughter Eléonore joined the company in 2023 as vice-president, having previously worked as a lawyer in Paris. Her enthusiasm for the career change is unforced: 'It's been a blast since I joined … I'm not passionate about law, there isn't fun every day … It's a nice change of scenery from Paris.'

In time she will likely head up the firm whose holding of more than 26 hectares on the Hill of Corton dwarfs all others. Those slopes provide the grapes for Latour's long-time standard-bearer, the Corton-Charlemagne, a wine of rich, refined fruit that deserves a decade's ageing, during which it develops a creamy, mouthfilling texture. The style harks back to a time before the current trend for early picking and lean acid, which, unless handled properly, can be shrill. This wine's renown is something of a two-edged sword, for it shows what Latour is capable of while also asking why there is a such a gulf between it and the *premiers crus* such as the Meursault 'Château de Blagny'.

The red wines that for decades failed to raise a smile on this taster's face are now more assured. The bland flavours of old have given way to a fuller profile, replete with crisp fruit and clear definition. What was once vapid now approaches vibrancy. After hand harvest into 15-kilogram boxes the grapes are sorted at the winery by a team comprising company retirees, who welcome the chance to chat about the past while contributing to the future. The fruit is mostly de-stemmed, though a layer of whole bunches is placed at the bottom of each vat. Vinification now lasts two to three weeks, longer than previously, with *pigeage* by foot to extract more colour. Élevage takes place in barrels made by Latour's own cooperage and the wines are flash-pasteurized before bottling. This latter practice is utilized to avoid filtration, though it has been the subject of prolonged debate over the years. Nevertheless, wines such as the *premiers crus* Volnay En Chevret and Pernand-Vergelesses Ile des Vergelesses are pointing the way to a more quality-focused future for the red wines.

Try this: Beaune 1er cru Les Perrières

There are many 'Perrières' vineyards in the Côte d'Or, the name indicating the presence of old stone quarries that pre-dated the vines. Here, at the northern end of the Beaune appellation, the soil is stony and poor, and the slope steep. The rugged origins are reflected in the wine which, if not quite rugged, boasts noticeable grip and substance; there's plenty of dark fruit stuffing but it stops short of being oppressive. With more power and structure than is normally associated with Beaune, this wine gives the lie to the commonly held perception of Beaune as a pleasant but light one-trick-pony.

Maison Benjamin Leroux

5 rue Colbert, 21200 Beaune
Tel: +33 3 80 22 71 06

'Winemaking is easy,' said Benjamin Leroux to a roomful of wine lovers on one of his regular visits to Ireland some years ago. That got their attention. Here was a noted Burgundian saying there was nothing to it, no need to rely on arcane practices handed down from generation to generation, or complex oenological theories to baffle neophytes. Specifically referring to white winemaking, he continued: 'You simply have to get the harvest date correct.' Today, when picking in the second half of August is no longer unusual, even a day one side or the other of ideal can be critical. A day's sunshine then will have a markedly greater ripening effect than a day's sunshine in late September. The target is smaller, the sweet spot has shrunk, greater diligence is needed.

'Diligence' could be Leroux's middle name, he's not one for complex elaborations of his winemaking philosophy and practices, he simply goes about his task methodically and with quiet assurance. It is a little over ten years since he left his position as winemaker at Comte Armand in Pommard to strike out on his own as a Beaune-based négociant, and that decade has been one of steady development and improvement, in terms both of product range and reputation. Not that he had anything to prove, his reputation was already high, it has simply been embellished in recent years. Progress has been by steady increments, a gradual accretion of assurance in the wines, of which there are dozens of appellations.

A cellar tasting with Leroux involves shuttling back and forth between rows of barrels as he draws a sample here and another there. A faint smile plays permanently on his face, the eyes twinkle and the voice is soft yet insistent, just like his wines. There are no shouty flavours, as his two iterations of the Meursault *premier cru* Genevrières

178 CÔTE D'OR

ably demonstrate: the *dessous* version, from the lower section of the *climat*, is abundant yet refined, while the upper, *dessus*, version is fuller and more vibrant, with notable intensity to carry the impressive weight of fruit.

Leroux's reds are firmer and more structured than the whites. The Volnay 1er cru Clos de la Caves des Ducs, a *monopole* of 0.64 hectares, may not have the gravitas of Volnay's better known *premiers crus* but it is no shrinking violet either, its floral fruit well supported by a foundation of civil tannin and tingling acidity. The Vosne-Romanée, of which he produces a relatively generous 20 barrels, boasts measured elegance in a structured framework, with ripe fruit to the fore and a lingering finish that belies its humble *village* status.

Concluding my profile in the first edition of this book I wrote: 'Leroux is currently producing excellent wines, but despite what he has achieved so far the best is yet to come.' To say now that he may yet do better still should be seen as an observation, not a criticism of progress to date.

Try this: Vosne-Romanée 1er cru Au-Dessus des Malconsorts

Barely more than one hectare in area, this vineyard lies above the highly regarded Malconsorts itself and, while it may not claim the same gustatory grandeur as 'big brother', in the hands of Benjamin Leroux it produces a remarkable wine. He makes one large barrel, equivalent to about one and a half *pièces*, of vividly-flavoured wine, full-bodied but not heavy. Succulent fruit is leavened by perky spice, with a vein of crisp acidity running in the background. The wine is deep and complex, revealing new nuances of flavour with each sip.

Domaine Seguin-Manuel

2 rue de l'Arquebuse, 21200 Beaune
www.seguin-manuel.com; tel: +33 3 80 21 50 42

As with other Côte d'Or domaines, recent expansion at Seguin-Manuel has been by way of new vineyards to the south in the Côte Chalonnaise and the Mâconnais, in this case Montagny, Givry and Pouilly-Vinzelles. Possibly the more significant news, however, was the adoption in 2022 of biodynamic practices in the vineyards, the wines having been certified organic since 2015. 'It made a big impact, immediately noticeable in 2023,' says Thibaut Marion, who has owned Seguin-Manuel, along with a partner, since 2004. 'The vines reacted better to weather and diseases, and the ground was more friable and aerated. I am pretty sure that part of it is working, though we can't explain everything. It may yet be proven.'

It is Marion's refreshing absence of any proselytizing on the subject that makes him a more convincing advocate for biodynamics than most. He passes on his observations without preaching and he is not in a hurry to seek certification: 'We will see how it works first.' Despite his advocacy, he readily concedes that if climate change continues apace biodynamics 'may not be strong enough to cope', acknowledging with a note of resignation, 'if it is too warm, it is too warm'. Regardless, he is not about to throw in the towel and is exploring the use of drought-resistant rootstocks, as well as pruning his vines higher than 10 years ago.

The red wines are the flag carriers at Seguin-Manuel and all are true to their origins, most particularly a spicy and forceful Pommard Petits Noizons, made with 40 per cent whole bunches, that might benefit from a little more refinement. Marion follows no set de-stemming regime, adjusting according to vineyard and vintage; the grapes are sorted at the winery using a vibrating table and conveyor belt before fermentation at about 30°C in stainless steel. 'We want a soft extraction,' he says, so *pigeage* is done by foot: '*Pigeage aux pieds*' he terms it, adding, 'each cuvée is tailor made.' That said, there is a noticeable house style, brisk fruit verging on brusque at times, as in the Nuits-Saint-Georges Vieilles Vignes. Regular cork closures are used at bottling, Marion reasoning that to use anything else, 'would be a nonsense, because we are organic.'

Parallel with the domaine wines is a range of wines from bought-in grapes, made in the same way and labelled almost identically, as 'Seguin-Manuel' compared with 'Domaine Seguin-Manuel'.

Try this: Vosne-Romanée Aux Communes
Thibaut Marion bought this 1.8-hectare plot of *village* Vosne-Romanée, his largest, in 2006 when Domaine René Engel was sold and renamed Domaine d'Eugénie. Previously, about 25 per cent whole bunches were used but today it is fully de-stemmed. 'Whole bunch doesn't bring anything to this wine,' say Marion of his flagship, whose vines are half-a-century old. There is a tingly, spicy intensity, though it is not heavy, it leans towards the fine and delicate end of the flavour spectrum. Its class is signalled by good length on the finish, not weight on the palate.

180 CÔTE D'OR

POMMARD

The village clusters tight around the four-square church with its immediately recognizable blocky tower. Time was when the wines mirrored its imposing countenance: commanding and structured though hardly charming. Sturdy wines made in homage to sturdy church.

This is solid red wine country – there is no appellation for white – and if the wines have received a little too much condemnation for coarseness and lack of charm, that shows signs of changing. Too often in the past the tannins and fruit stood separate, refusing to engage, and the flavour oscillated between those two poles. Grace came with age but sometimes reluctantly. Wines of that ilk can still be found but increasingly a good example boasts a more harmonious flavour than heretofore. Pommard may have big feet but in the right hands they can be neatly shod. The winemakers now vinify with a gentler hand; it is as if some of Volnay's charm has seeped northwards, replacing fortitude with finesse.

The compact village divides the roughly square commune in two, with a block of *premier cru* vineyards to north and south, themselves in turn flanked by *village* vineyards on the lower land beside the main road and on the hillsides above the village, with a tail of roughly south-facing vineyards along one side of the Grande Combe. Through the combe flows the Avant Dheune, little more than a stream that divides the village, hidden from view in the centre though clearly visible if you approach along the tiny rue Marey Monge. Just as Volnay and Chambolle are often twinned in burgundy lovers' minds, so too are Pommard and Nuits-Saint-Georges and it's not hard to see why. Both are muscular renderings of Pinot Noir and, as with Nuits, Pommard has enjoyed easy recognition – apart from the silent 'd' there are no pronunciation pitfalls, unlike Montrachet for instance, and its proximity to Beaune makes it accessible for visitors to the region.

Producers

Domaine du Comte Armand – Clos des Epeneaux

7 rue de la Mairie, 21630 Pommard

www.domaine-comte-armand.com; tel: +33 3 80 24 70 50

The Armand family have owned the Clos des Epeneaux for almost 200 years and until 1994 owned no other vineyards, so that 'Armand' and 'Epeneaux' are often used interchangeably when speaking of this domaine. One of the 1994 purchases included a hectare of Auxey-Duresses *premier cru* that today makes an utterly beguiling wine, pure and long, with a mild rasp of cherry fruit supported by a streak of crisp acidity.

As with all the wines, it is 100 per cent de-stemmed, the grapes being cooled overnight before transfer of mostly intact berries to stainless steel for a week's cool maceration at about 10–12°C. After six days the cooling ceases, allowing fermentation to start. Winemaker Paul Zinetti then employs a combination of two to three punch-downs per day early on, changing to one pump-over per day later in the fermentation.

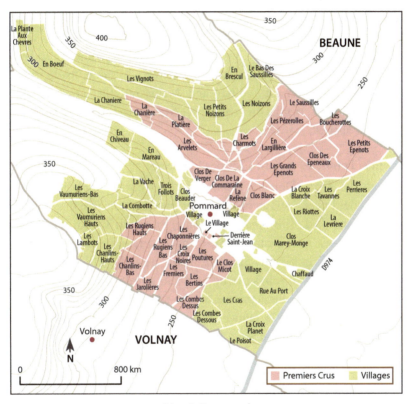

Map 8: Pommard

Thereafter, a further maceration follows, when he likes to keep the cap wet, ever mindful that with Epeneaux: 'It is very easy to make a monster, it is important to be careful with it.' Time in barrel, about 30 per cent new for the *clos*, will cover two winters, 18–24 months, before bottling, with natural corks used for all the wines: 'I prefer to pay a big price for good corks,' says Zinetti.

182   CÔTE D'OR

Zinetti describes the *clos* as, 'a big baby' and in recent years has utilized a flock of eight sheep for four months in winter to nibble between the vines, fertilizing as they go. He also notes that it has its own microclimate, thanks to the reflected heat from the big wall, the nearby vines growing quicker and ripening earlier. This was undoubtedly a benefit in days gone by, but cuts the other way now; as harvest approached in 2020 the potential alcohol surged upwards by 1.5 percentage points in just two days. Regardless of this contemporary challenge, Zinetti is crafting wines that steer clear of hearty flavours that pummel the palate. After tasting across a range of wines from several vintages, my abiding impression, noted at the time, was of, 'walking away with lovely freshness in the mouth.'

Finally, as with so many prestigious Côte d'Or producers today, there is also an Aligoté of real style and charm. This one, made from three plots in Pommard, Volnay and Meursault, the latter having 90-year-old vines, majors on green apple fruit and a mild mineral bite. A perfect aperitif.

Try this: Pommard 1er cru Clos des Epeneaux

Though the Auxey-Duresses *premier cru* cries out for inclusion here, the Clos des Epeneaux is so synonymous with this domaine that it cannot be overlooked. This *monopole* lies almost wholly within the Petits Epenots *lieu-dit*, with a corner of the upper part, where there is only a scraping of soil, in Grands Epenots. In the past, the wine challenged rather than charmed: robust, structured and a little unforgiving, until a long stretch in the cellar tamed it for the table. Today, the structure is still there, now cloaked in refined fruit and civil tannins. The lingering impression is one of freshness and vibrancy.

Château de Pommard

15 rue Marey Monge, 21630 Pommard

www.chateaudepommard.com; tel: +33 3 80 22 12 59

Turning off the D974 to approach Château de Pommard takes you between high walls, graveyard to left, vineyard to right. That vineyard is the Clos Marey Monge – essentially the château's front garden – and if a prize was given for the most cossetted and pampered *village* vineyard in the Côte d'Or it would be a strong contender. The château was purchased by Michael Baum from the United States in 2014 and the decade since has seen a level of renovation and renewal that was probably the equal of all such work done in the previous three centuries. At time

THE VILLAGES AND PRODUCERS OF THE CÔTE DE BEAUNE 183

of writing it was still ongoing; hopefully the works will have finished in time for the tercentenary of the *clos* in 2026.

The cosseting of the *clos* began with significant replanting and conversion to organic and then biodynamic viticulture, gaining certification in 2020. A comprehensive soil analysis was also undertaken, identifying and naming seven different sub-terroirs: Chantrerie, Émilie, Grands Esprits, Nicolas-Joseph, 75 Rangs, Vivant Micault and Simone. A different cuvée is produced from each of these and, while it is an interesting exercise to taste the stylistic variations between them, it is like the principal players in an orchestra being tasked to perform as soloists, and then together as an ensemble – in this case the Clos-Marey Monge, which is blended from all seven. Ultimately, the ensemble delivers a more complete, nuanced expression of what the entire vineyard is capable of.

Grapes were previously 100 per cent de-stemmed but today about 30 per cent whole bunches are used by winemaker Paul Negrerie, who was appointed in 2021. He brings worldwide experience to the role, having previously worked in New Zealand, South Africa, Australia, Israel and the United States. With the 2022 vintage he signalled a move towards greater refinement in the wines, more perceptible in texture than flavour, resulting in a more polished profile.

After hand harvesting and sorting on a vibrating table, fermentation takes place in open-top stainless-steel tanks with *pigeage* initially by foot. Once fermentation is complete the wine is pressed using a vertical press, with all the press wine used, and then élevage takes place in standard 228-litre *pièces* and two amphorae of 500 litres each for 15 months before bottling after a light filtration.

Much has been achieved at Château de Pommard in recent years. Might another achievement follow by way of elevation to *premier cru* status for all or part of the Clos Marey-Monge? Perhaps the half-hectare plot Simone, which lies tucked into the north-west corner of the *clos*, between château and village, yielding a wine of compelling class, would be the most likely candidate. Given the glacial pace of such changes it might be the quatercentenary before this happens.

Try this: Pommard Clos Marey-Monge
The 20-hectare *clos* stretches a couple of hundred metres across flat ground from the château to the D974 and the recent refining and lightening in this wine's style is signalled before it is even tasted – by the change in bottle shape, from the heavy cudgel of old to a slim skittle more in tune

with today's zeitgeist. A smooth tannic base supports lighter red fruit and floral flavours, with some mild spice adding vigour on the finish. Michael Baum describes his wine: 'It has the power of Pommard but the finesse of Volnay, I often think of it as a marriage between those two villages.'

VOLNAY

Few villages catch the eye like Volnay thanks to its being set a little up-hill, with nothing to obscure the view from the road; it sits tucked into a fold in the hillside and invites the cliché 'sleepy'. Volnay is a prime example of the characteristic coalescing of vineyard and village seen repeatedly

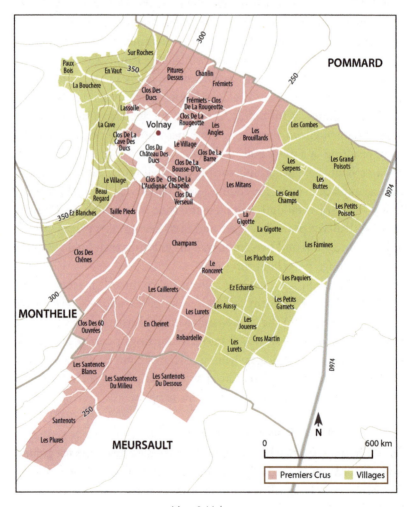

Map 9: Volnay

along the Côte d'Or. Do the village streets thread their way between the vineyards, or do the vineyards insinuate themselves between the streets? It is not possible to say where one ends and the other begins; they simply segue into each other. There's history in Volnay, courtesy of long ducal and royal connections that are reflected in the vineyard names: Clos des Ducs, Clos de la Cave des Ducs and Clos du Château des Ducs. The *premiers crus* straddle the D973 road as it bisects the commune, rising towards Monthélie and Meursault, and affording an excellent view across the vineyards and the Saône plain, and on towards Mont Blanc.

The name Volnay is uncomplicated, bereft of the hyphenated clutter that makes many others a mouthful. It's straight and direct, mirroring the immediate appeal of the wines and the fact that they are all red; no colour qualification needs to be added. Traditionally the wines were regarded as elegant and fine, all feminine charm to Pommard's masculine heft. Today's archetype has filled out without adding overt weight, there's more depth and darker fruit than before and the wines are stronger, but ideally it should be revealed in intensity over concentration. They still have an insistent appeal; tingling with vinous energy they are compelling expressions of Pinot Noir at its most seductive. It would be wrong, however, to dismiss twenty-first century Volnay as charming and little else; the attraction is enduring thanks to the intensity of fruit flavours and the skill of the quality-conscious vignerons.

Producers

Domaine Marquis d'Angerville

Clos des Ducs, 4 rue de Mont, 21190 Volnay

www.domainedangerville.fr; tel: +33 3 80 21 61 75

It was only when I mentioned my love of good Aligoté that Guillaume d'Angerville disappeared into his cellar to add a bottle to the largely red line-up of wines for tasting on my most recent visit. From the 2022 vintage it was a delight, racy and poised, with lovely tension and a stony, lemon-pith aftertaste. Remarkably, d'Angerville's father, Jacques, had given away a plot of Aligoté to a local farmer years before, only for the farmer to return it more recently. 'We keep the sharpness. It sells like hot cakes,' says d'Angerville. Not surprisingly.

This, however, is a Volnay house through and through whose reputation rests on its enviable roster of eight *premiers crus*, totalling 11.5 hectares out of the domaine's 16. Included in their number is the two-hectare *monopole* Clos des Ducs that forms the most prized back

186 CÔTE D'OR

garden in Volnay. Though its origins can be traced back over 500 years, the domaine as constituted today dates from 1906, when it was bought by Marquis Jacques d'Angerville, a champion of domaine bottling in the 1930s. His grandson, Guillaume, took over in 2003, having first pursued a career as a banker with JP Morgan, and today his daughter Margot has joined him with a view to eventually succeeding him.

Guillaume d'Angerville is quietly spoken but articulate; when he speaks it is worth listening, particularly about climate change. He is convinced that the vine is adapting, particularly to recent roasting summers that in the past would have branded the wines with unpleasant, singed fruit flavours. For him, the 2018 season represented a turning point, with fermentation difficulties caused by, 'tired yeasts'. By 2020 he noticed that the vines were, 'less impacted by drought and heat,' even though conditions were much the same and, 'since then I am regaining hope that Pinot Noir will be what we want it to be.' He also credits his conversion to biodynamic viticulture, started in 2006 and completed in 2009, with helping the vineyards to cope better with the torrid conditions. In all of this he is careful to clarify that he is not making a dogmatic claim, simply an observation based on anecdotal evidence.

Replanting is a subject that is also on d'Angerville's mind, and he notes that his grandfather made the then sensible decision to plant the vineyards with 1.2 metres between rows and 0.8 between vines to get more sun exposure. That layout is not ideal today but to change it across the whole estate would be a huge undertaking that, 'would take several generations. It's not an easy one to solve,' though he seeks to mitigate the difficulties caused by the historic plantings by pruning later and training the vines higher.

Lest this sound like a litany of unrewarded challenge, followers of this domaine can rest assured that the wines, made from 100 per cent de-stemmed fruit and given a cool pre-fermentation maceration, are still models of their kind. Élevage takes place in barrel, about 20 per cent new, to deliver a range of wines that rely on subtle insistence rather than concentration to make a lasting impression. There is no overt weight, no harshness needing a decade in the cellar to be tamed, only impressive harmony and balance.

Try this: Volnay 1er cru En Champans

D'Angerville is easily the biggest owner in Champans, with 4 hectares out of the vineyard's 11. The name possibly derives from *cham*, meaning 'field' and *pans* possibly refers to the slope, which is gentle here, as is the wine. This is *vrai* Volnay, guaranteed to win over any who are new to the village's charms. A rush of juicy sweet fruit might descend into lushness were it not for the tingling acidity that holds it in check. Singing length on the finish immediately leads to another sip.

Domaine Michel Lafarge

15 rue de la Combe, 21190 Volnay

www.domainelafarge.com; tel: +33 3 80 21 61 61

As with many Burgundian domaines, a satnav that works underground would be a good accessory when visiting the Lafarge cellars, a series of interconnected spaces that twist and turn over several levels. The oldest section – arched and mould-dappled – dates from the thirteenth century and rubs shoulders with twentieth-century additions from 1974 and 1987. For many decades the patriarch here was the eponymous Michel, a true gentle man of wine, held in universal esteem and affection, who died in 2020. Today, his son, Frédéric, and granddaughters, Clothilde and Eléonore, continue his work in the same unobtrusive, far-from-attention-seeking way.

Though the vineyard holdings include parcels in Beaune, Pommard and Meursault, Lafarge is essentially a Volnay domaine, where the impressive range of *premiers crus* includes the *monopole* Clos du Château des Ducs that could be regarded as the Lafarge back garden. The vineyard covers a little over half a hectare and the name refers to the château built by the dukes of Burgundy in Volnay in the eleventh century. Thanks to heat reflected off the walls, its 30-, 50- and 70-year-old vines ripen early and it can be harvested in a single morning. It is the domaine's favoured child, receiving kid-glove treatment by way of de-stemming that is done by hand. Since 2016 a team of five or six roll the grapes back and forth across open-weave wicker 'sieves' placed on top of the tanks, allowing the berries to drop through slowly. 'It is very soft and gentle, it gives purity and energy, and creates a calm ambience at the beginning of harvest,' says Eléonore Lafarge.

The Caillerets is also worthy of note, for the fact that, since 2020, following an idea that Frédéric always wanted to pursue, a section of the Lafarge parcel there has its vines planted in a spiral pattern, taking inspiration from South American agriculture. The lines follow the shape of the plot and each vine has a slightly different exposure from its neighbour. The first vintage was 2023 and it was blended with the

188 CÔTE D'OR

conventional section. Whether it has a perceptible effect on the wine's quality remains to be seen.

A friend once referred to Lafarge wines as the Barolos of Burgundy, an apt description. These are wines that can show beautifully when young but which then shut down for years; they can be dumb and even disappointing, coiled tight and remarkably uncommunicative. They are not for the impatient; they are built for long-haul ageing. Time, measured by the decade rather than the year, is needed for these wines to sing in 'full throated ease'.

Try this: L'Exception, Bourgogne Passetoutgrain

This wine is aptly named, for it challenges accepted practices at many steps of its production. The 1-hectare vineyard lies close to Meursault and is co-planted with equal parts of Pinot Noir and Gamay vines that are now nearly 100 years old. 'They synchronize,' says Eléonore Lafarge when explaining that the grapes ripen at the same time, so can be harvested together before fermentation together. This is not a complex wine; it is simple in the best sense. There is a direct hit of immediately appealing ripe red berry fruit, fresh and light, enlivened by a discreet sharpness, and then it is gone, with no long farewell on the finish.

Domaine de la Pousse d'Or

8 rue de la Chapelle, 21190 Volnay

www.lapoussedor.fr; tel: +33 3 80 21 61 33

The grand façade of Pousse d'Or looks out over Volnay's lower slopes and is clearly visible from the main road, making it the easiest address to find in Volnay, especially when lit by the early morning sun. For some 30 years at the end of the last century the domaine enjoyed a high reputation thanks to the efforts of Gérard Potel, a winemaker of renown who died suddenly in 1997, just before Pousse d'Or was acquired by Patrick Landanger. The early years of his tenure, when a manager ran the domaine on his behalf, were not so successful. Customers drifted away until he took charge himself and got things back on a sure footing – today's declaration on the website's homepage, 'Refinement in a bottle,' is accurate and not the sometimes-vacuous puff that pollutes many websites. Pousse d'Or is now numbered amongst Volnay's best again.

More recently there has been generational change; Landanger's son, Benoît took over from his father in 2018 and today is in sole charge, Patrick having died in 2023. Though firmly rooted in Volnay,

Pousse d'Or also has a notable footprint in the Côte de Nuits, courtesy of holdings in *grands crus* Bonnes Mares and Clos de la Roche, but most particularly a trio of Chambolle-Musigny *premiers crus*, that includes a slice of Les Amoureuses. As to be expected, the Amoureuses is truly special, a *grand cru* in all but name that is not shamed when tasted alongside the Bonnes Mares, though the négociant Charmes-Chambertin leaves it a little in its wake, thanks to a surging intensity of flavour, dark and deep.

Red wines account for almost all of Pousse d'Or's production. After hand harvesting and a preliminary sorting in the vineyard the fruit is all de-stemmed before passing through an optical sorter, the intact berries then descending by gravity to the tanks below. Great refinement is brought to the sorting process – any grapes picked in the afternoon are cooled overnight to 10°C in a new cold room first used for the 2022 harvest, the process being more efficient when the grapes are cool. Fermentation is all in stainless steel, save for two wooden vats used for the *monopoles* Clos de la Bousse d'Or and Clos des 60 Ouvrées. There are two punch-downs per day, with limited pumping over. During fermentation the previous year's wine is racked out of barrel and into stainless steel for 4–6 months before bottling in February or March, thus freeing up the barrels for the current vintage, which only remain empty for the minimum of time. About 20 per cent new oak is used, down from the 30 that was more usual up to 2016.

Pousse d'Or has also been using amphorae for a portion of three of its Volnay *premiers crus*: Bousse d'Or, 60 Ouvrées and En Caillerets. These are bottled separately from the 'regular' wines and provide an interesting comparison with them. The amphorae bring a transparency to the flavour, like a lighter weave in a fabric; the fruit is foremost with the richer elements that characterize a traditional élevage largely absent. Regardless of the vessel used, the Pousse d'Or style is structured yet supple, the tannins are discreet, the balance is good, save for the Pommard which kicks against the traces of the refined house style, lacking the elegance of its Volnay neighbours.

It is hard not to regard the white wines as footnote here, but they should not be overlooked. The Puligny-Montrachet 1er cru Les Caillerets comes replete with a cornucopia of fruit, from pineapple to citrus to green apple, while the Chevalier-Montrachet, from a 0.12-hectare plot, is more reserved, only displaying its depth and intensity after repeated swirling, sniffing and sipping.

190 CÔTE D'OR

Try this: Santenay 1er cru Clos de Tavannes
Pousse d'Or owns two hectares of Tavannes, which is a five-hectare *lieu-dit* within the Gravières vineyard. It sits right on the boundary with Chassagne-Montrachet and the name comes from previous owners, the de Saulx-Tavannes family. Many regard Tavannes as the best red of Santenay and tasting Pousse d'Or's it is not hard to see why. There is a depth and conviction of flavour that might place it in the Côte de Nuits – and it can age superbly. The 1970, drunk on several occasions at 40-plus years old, was magnificent: scented, satin-textured and singing on the finish.

MONTHELIE, AUXEY-DURESSES AND SAINT-ROMAIN

It would be overstating the case to describe these three villages as the forgotten trio of the Côte de Beaune; overlooked and hence underrated is more accurate. They form part of the inner or secondary line of vill-ages that distinguishes the Côte de Beaune from the Côte de Nuits with its more regimented parade. Geography has conspired against them. You must divert away from the well-trodden vineyards of Volnay and Meursault to reach them and they are not on the road to another vinous destination. Saint-Romain sits cheek by jowl with steep rock face and as you travel towards the village there is a feeling of separation from the main côte, a sense of dislocation and isolation engendered by the forest-ed hills that rise steeply from the roadside. It is more rural here; there's none of the smug comfort that characterizes patrician Meursault barely four kilometres away.

In the case of Auxey-Duresses and Monthélie some of their vineyards abut those of Meursault and Volnay, and produce similarly satisfying wines. A well-made Auxey-Duresses Les Hautés is a delight and com-pares favourably with good Meursault; equally, Monthélie 1er cru Sur la Velle, next door neighbour to Volnay 1er cru Clos des Chênes, can possess the same charm as its more storied neighbour.

It is in villages such as this trio that the surge in quality in recent dec-ades has been most pronounced, so that now, more than ever, it behoves the Burgundy lover – not the collector or investor – to explore these 'lesser' communes of the Côte d'Or. As with Saint-Aubin, since the turn of the century the wines are gradually gaining recognition beyond the cohort of savvy buyers who have been following the good producers for years. If the ever-spiralling prices for the famed *crus* cause despair, it is from villages such as these that quality wines at non-ludicrous prices will be found.

Producers

Domaine Henri & Gilles Buisson
Impasse du Clou, 21190 Saint-Romain
www.domaine-buisson.com; tel: +33 3 80 21 22 22

The Buisson family can trace its origins in Saint-Romain back to the time of the Crusades, though the domaine as currently constituted is considerably younger, being founded by Henri Buisson in 1947. His son Gilles and Gilles' wife Monica expanded their holdings to 20 hectares in the latter years of the last century: 12 hectares in Saint-Romain, with the remainder scattered across the Côte de Beaune, including some choice parcels on the Hill of Corton, split evenly between red and white and encompassing about two dozen appellations. Today the baton has passed to their sons, Franck and Frédérick. The former, with a business degree, looks after the commercial side and the latter, with a diploma in oenology, takes care of the winemaking.

A visit to Buisson strays pleasurably far from the usual routine of tasting followed by winery tour. Franck loads up the 4x4 with bottles and glasses and takes his visitor on a bouncing vineyard drive, all the while elaborating on the brothers' application of biodynamic principles to their viticulture. (Whatever your view on biodynamics Franck Buisson is the man to listen to if you want a good, hokum-free exposition of it.) Most pertinently he emphasizes the need for extra manpower and equipment, and to recruit workers who are, 'believers,' and cognizant of the thinking behind the practices.

The ultimate test of those practices is, of course, the wine in the glass, otherwise their labour-intensive application is little better than putting go-faster stripes on an old banger of a motor car. Without a dextrous guiding hand it is a waste of time and effort – neither is being wasted at Buisson.

All the grapes are hand-picked into 20-kilogram boxes and taken on a small flatbed truck to the winery in less than 30 minutes, where they are sorted on two tables, one vibrating, manned by 10 people. For the white wines the grapes are crushed and pressed, with some maceration in the press, the juice being sent directly to barrel. As Franck Buisson

192 CÔTE D'OR

puts it, 'the juice is thick and dark, with texture and deepness'. Various barrel sizes are used, 228, 350 and 500 litre, as well as large foudres, and the brothers don't do much bâtonnage, perhaps three or four stirs, no more.

For the reds the de-stemming regime is flexible, never 100 per cent either way, before vinification in concrete tanks. *Remontage* and *pigeage* are done as needed: 'There is no recipe, we taste the fermenting wine every day and decide what it needs, to find the balance of every terroir.'

Try this: Saint-Romain Sous le Château

Driving into Saint-Aubin from Auxey-Duresses the Sous le Château vineyard rises steeply up to 370 metres on your right, the name referencing a long destroyed ancient fortress. The exposure swings from east to south, the thin soil resting on hard limestone. The wine displays counterpoised flavours of peach and citrus, jostling on the palate and all wrapped in a glossy texture, mention of which prompts Buisson to talk of its 'touch in the mouth'. That touch goes on long and smooth into an echoing finish.

Domaine Alain Gras

rue Sous la Velle, Village haut, 21190 Saint-Romain
www.domaine-alain-gras.com; tel: +33 3 80 21 27 83

After it dog-legs out of the village of Saint-Romain the Rue du Grand Tour rises sharply and narrows alarmingly, with forest closing in on both sides for a couple of hundred metres before the ground flattens and opens at Domaine Alain Gras. There has been much change here since the first edition of this book, most visibly in the new winery that saw its first vintage in 2023. Alain Gras, who has been in charge here since 1979, has now been joined by his son, Arthur, who started working alongside *père* in 2020. There is also an impressive new tasting room that can accommodate up to 20 visitors, complete with a forbidding wooden effigy of Saint Vincent to look over them.

What has not changed is the consistent quality of the wines – harvested from a total of around 15 hectares, 13 of which are in Saint-Romain with about 1.5 each in Auxey-Duresses and Meursault. Production divides 70:30 in favour of white wines, which are models of balance and harmony. They satisfy rather than challenge the palate and they keep on satisfying, sip after sip. Lightness and freshness are their calling cards; they don't try to be something they are not. Élevage takes place in a combination of barrels from near neighbour François Frères and stainless steel tank, with assemblage and bottling just before the following harvest. The proportion of new oak rises from about 30 per cent for the Saint-Romain, through 40 for the Auxey-Duresses, to 50 for the Meursault.

Several factors influence the Gras style, principally the vineyards' altitude of about 400 metres and the fact that the commune occupies a side valley, resulting in a cooler microclimate. The altitude also means that there is little more than a 20-centimetre veneer of topsoil above the limestone, endowing the wines with a crisp mineral hit. Also, Gras makes a single Saint-Romain cuvée in white and another in red, blended across his holdings in the commune. This contrasts with some of his neighbours who now vinify and bottle several *lieux-dits* separately.

The red wines are made with 20 per cent whole bunches and undergo a similar élevage. Today they include some wines, such as a dark and deep Pommard 1er cru Clos de Verger, that Arthur makes in his own name from bought-in grapes. He is developing a small but expanding négoce business that produced 5,000 bottles in 2022 and 15,000 in 2023. All this happens in the new winery, which has allowed the whole vinification process to be rationalized, bringing everything together under one roof where previously a scatter of locations had to suffice. The future is in safe hands at Domaine Alain Gras.

Try this: Auxey-Duresses Très Vieilles Vignes

The wine is made from a hectare of vines planted in 1904 and still in good condition, which produce an impressive yield of 20 barrels per year on average. This is no pensioner on the palate, thanks to a vigorous structure and vivid flavour that belie its humble *village* status. The tannins are grippy and the fruit has a pleasant rasp, with a mild *sauvage* character and a spicy kick on the resounding finish. For greatest appreciation do not ignore this critical recommendation from the domaine: 'This wine can be served a bit cool but not at room temperature.'

Maison Frédéric Leprince

7 Grande Rue, 21190 Nantoux
www.frederic-leprince.com; contact@frederic-leprince.com

Vinous hair splitters might cavil at the inclusion of a winemaker based outside the Côte d'Or but, given that Frédéric Leprince's range includes the likes of Vosne-Romanée and Gevrey-Chambertin, he is worthy of

194   CÔTE D'OR

his place. And the several cases of his excellent Aligoté, bought and drunk with great pleasure in recent years, won the argument for him.

Leprince hails from Normandy but got the wine bug young, thanks to sharing memorable bottles with a wine-loving uncle. A wandering wine education followed, in Beaujolais and Bordeaux, before he settled in Burgundy, working with Etienne de Montille for a year at Château de Puligny and a further year at Comte Armand, with Benjamin Leroux, who he describes as 'a magician of Pinot Noir'. He made his first vintage in his own name in 2007, afterwards buying an old farmhouse in Nantoux, where he is now based. There's an antique feel to the place, from the ancient, rusted bell at the entrance to the sagging stonework of the courtyard within.

The wines run counter to this impression – there is no dilapidation in their energetic flavours, quite the opposite. Production splits a little in favour of white wines, 55:45, the grapes for which are pressed immediately after hand picking and the juice left to settle for 24 hours, prior to a long fermentation that can last several months in 600-litre barrels.

Leprince elaborates in greater detail when talking about his reds. The grapes are sorted on a vibrating table, 'I only keep grapes I would eat', he adds. Vinification, including some whole bunches, is in wooden vats, maceration and fermentation lasting for about three weeks, with *pigeage* by foot. The wine is then transferred to barrels in a variety of sizes, including 225-litre Bordeaux barrels as well as 350- and 500-litre vessels, from Taransaud, Billon and Eric Millard. Leprince describes the effect of these latter, hard-to-source, barrels as, 'like cream', and manages to buy ten every year. He is also hugely proud of his vertical press: 'It is the best thing I ever bought,' he says while patting it like a favoured pet.

Aside from the Aligoté, the red wines carry more assurance than the whites, seen to good effect in the Beaune 1er cru Les Cents Vignes, which is made with 40 per cent whole bunches and delivers a richer, fuller palate of dark fruit than one usually associates with Beaune. Leprince is modest about his achievements so far: 'I am sure I can do better, especially in the reds. I am at 65 per cent of the quality that I want. I want to do better.'

Try this: Volnay

Blended from a trio of contrasting vineyards above and below the village: En Vaut lies above Clos des Ducs and rises to 375 metres at the treeline, while Les Grands Champs and La Gigotte are over 100 metres lower, on the flat below Les Mitans. The result, from 85-year-old vines, is a wine that cleaves to the commonly held perception of Volnay: grace before structure, elegance before concentration. The texture is smooth and the finish is satisfying. The downside? Only about two barrels are produced.

Domaine Pierre Vincent

2 chemin sous la Velle, 21190 Auxey-Duresses

www.domainepierrevincent.com; tel: +33 3 80 22 80 31

After decades spent burnishing other people's reputations, first at Domaine de la Vougeraie and more recently Domaine Leflaive, Pierre Vincent has now struck out on his own, with the help of backers Hervé Kratiroff and Eric Versini. Together they bought the well regarded 7-hectare Domaine des Terres de Velle in Auxey-Duresses in 2023. Initially, Vincent divided his time between the new venture and Leflaive, before leaving the latter completely at the end of 2024.

Though Terres de Velle was founded as recently as 2009 the vineyard holdings are choice: in white, a slice of *grand cru* in Corton-Charlemagne, plus a slew of *premiers crus* in Chassagne-Montrachet, Puligny-Montrachet, Meursault and Savigny-lès-Beaune, while in red there are *premiers crus* in Volnay and Monthelie. They divide roughly into 5 hectares of white and 2 of red, spread across 20 appellations. Vincent likes working in miniature as it were, happy to have only three or four barrels of different cuvées, thus a tasting involves criss-crossing the cellar several times as he searches out wines to sample from the 2023 vintage, his first here. The cellar is notably humid, damp even, leading to less evaporation of the wines and reducing the need for topping up.

It is barely an exaggeration to say that Vincent is like a child on Christmas Day as he shows visitors around, elaborating on the philosophy and practices that he is applying to this new venture. He repeatedly points out that his biggest asset is the remarkably high average age of his vines – 58 years across the domaine, with the oldest approaching their centenary. 'My challenge is how to preserve the old vines,' says Vincent, making it sound like a mission statement.

While it is too early to identify a house style or comment on Vincent's methods, it can be safely said that, given his track record, he will turn out wines of finesse rather than richness, intensity rather than concentration. One gets a sense that despite his stellar career to date Vincent feels he is only starting, only now able to roll up his sleeves and do

196 CÔTE D'OR

things his way without a metaphorical glance over the shoulder to check he's following company policy. Previously, this was a slightly below the radar producer but that is about to change. It is early days yet but this is a domaine to watch.

Try this: Puligny-Montrachet

As might be expected from the man responsible for winemaking at possibly Puligny's greatest domaine for eight years, Vincent is no slouch when it comes to producing an elegant and beautifully balanced Puligny in his own name. It is sourced from two *lieux-dits*, one of which, Les Nosroyes, close to Clos de la Mouchère, has vines dating back to 1951. Referring to the eight barrels that he makes Vincent says, 'for me this is a large cuvée.' White flowers on the nose presage a delicate, lacy palate marked by harmony and persistence rather than outright power.

MEURSAULT

The village of Meursault possesses a scale and grandeur relative to its neighbours that is echoed by Gevrey-Chambertin in the Côte de Nuits. And Meursault the wine might be seen, without too ambitious a stretch of the imagination, as the southern white echo of Gevrey's reds. Their wines occupy similar positions in the firmament; they are bold, structured expressions of Chardonnay and Pinot Noir, seldom rivalled as statements of each grape's character. When one thinks of Meursault ethereal finesse is not the first quality that comes to mind. The classic descriptors utilize fuller flavour triggers: toast and honey, nuts and butter.

Travelling south on the main road past Volnay, Meursault is the first of the Côte de Beaune's triumvirate of great white wine villages you meet and is easily spotted thanks to the tall spire of the church that sits on a small rise at the centre of the village, across from the *mairie*. The village stretches away and down from the centre in different directions rather than packing tightly around it, and some substantial properties can be glimpsed above high walls and behind closed gates. The most impressive of them, however, the resurgent Château de Meursault, can be seen easily across its Clos du Château vineyard, especially when lit at night. The village sits towards the northern end of the commune and divides the vineyards, with the bulk, including the prestigious *premiers crus*, stretching south to Puligny. There may be no *grands crus* here but

THE VILLAGES AND PRODUCERS OF THE CÔTE DE BEAUNE 197

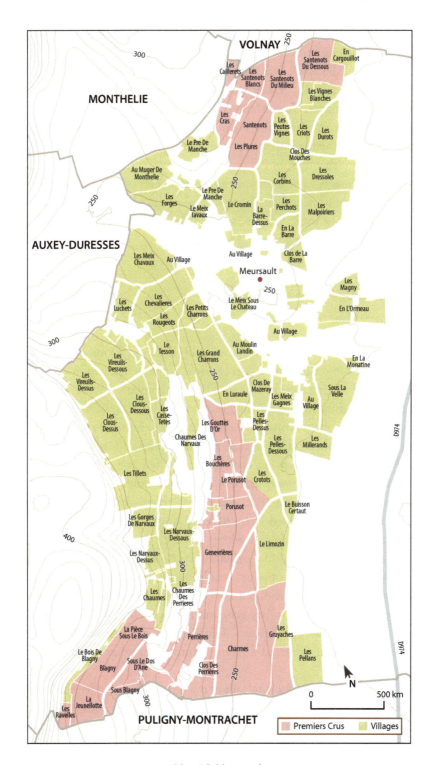

Map 10: Meursault

198 CÔTE D'OR

there's money in Meursault; the feel is solid and comfortable, and maybe a little self-satisfied.

Old style Meursault was full and forceful, a robust iteration of Chardonnay's qualities. Today's fashion pendulum, however, is swinging against this style. Thus Meursault now marches to many beats and not just a classic rhythm. Some are made linear rather than broad, tightly focused and intense rather than concentrated and rich. Some are not easily distinguishable from Puligny, and one hopes they have not been made in its image. Meursault has enough character of its own and does not need to ape its neighbour, however famed and sought after its wines may be. Without some fullness, some 'ballast' in the flavour the wine may be excellent but it won't be Meursault.

Producers

Domaine Robert Ampeau et Fils

6 rue de Cromin, 21190 Meursault

Tel: +33 3 80 21 20 35

Some places are different and some are really different. Domaine Robert Ampeau falls into the latter category. Most domaine visits follow a regular cadence, fitting loosely into a well-worn template. There's swirling and sniffing, slurping and spitting, interspersed with a routine question and answer that centres on a predictable roster of enquiries about planting density, whole bunches or not, use of oak and so forth. All that is forgotten at Ampeau, where a tasting is conducted in near total silence. Easy silence, not in any way unfriendly or forbidding.

I first came across Ampeau wines over thirty years ago and was not particularly impressed but, chancing upon a stash more recently that included a number of whites from the early 1990s – before the drab spectre of premature oxidation corrupted countless examples – I bought as widely as budget would allow. Included were several Meursaults, each wonderful, but none quite so splendid as the *premier cru* Les Perrières. It was lavish, generous and opulent, yet still focused, with no loose or sloppy flavours. Glorious succulence, with hints of honey and truffle, delivered a telling reminder of what has been lost through premox.

These wines were in the forefront of my mind as I descended the long flight of steps deep into the Ampeau cellar with Robert's son, Michel, who worked his first vintage here in 1970. Labelless bottles, sporting the bare minimum of identification, perhaps some small numbers in chalk, sat atop an upturned barrel. The surroundings were utilitarian, with steel cages housing thousands of bottles of back vintages towering on all sides.

The wines sang a far more exciting tune. The red Auxey-Duresses 2001 was vigorous and youthful, while the Volnay Santenots 2002 showed more age in a paler colour and softer flavours of spice and aged fruit. The real excitement was saved for the whites, however. A pair of Meursault *premiers crus* – Les Perrières 1997 and La Pièce sous le Bois 1994 – delivered a sensual flow of flavours that defied easy capture in words. Both were deep canary yellow, almost luminous, the Perrières suave and sumptuous, the Pièce redolent of mature marmalade. In terms of memorable, penetrating flavours, however, they were outclassed by the Puligny-Montrachet 1er cru Les Combettes 1995 (below).

Ampeau wines buck every trend, marching to their own beat and following no preordained path. They have an antique feel, a texture burnished by the decades; especially the whites, where the sensual flow of flavours calls for recalibration of the palate to assess them properly. Will we ever taste white Burgundies like this again? I fear not. It is a forgotten flavour, to the extent that today people coming across it for the first time may dismiss the wines as faulty. And faulty they are not.

Try this: Puligny-Montrachet 1er cru Les Combettes

The Combettes vineyard of nearly 7 hectares lies on the Meursault boundary, contiguous with Charmes, and in Ampeau's hands it might be considered a hybrid of the two communes, combining Puligny's finesse with Meursault's plumper profile. This wine is almost ageless; the 1995, drunk in 2023, was still restrained and intense on first taste. Only after repeated swirling did the central citrus core recede a little to reveal other nuances on the palate, flavours of honey and textures of satin. Wines such as this defy analytical analysis, far better to revel in their sensual delight.

Domaine Bitouzet-Prieur

7 impasse des Lamponnes, 21190 Meursault

Tel: +33 3 80 21 62 13

Though most people think of Bitouzet-Prieur as a Volnay domaine, its base of operations has been in new and expansive premises on the outskirts of Meursault since 2020. (This mirrors a move by other vignerons

200 CÔTE D'OR

in, for instance, Gevrey-Chambertin and Chassagne-Montrachet to move to custom-built premises away from cramped cellars in the heart of their village.) The new premises comprise two wineries, one for red and one for white, running parallel and bringing new emphasis to the term 'spick and span'.

Since 2007 the domaine has been in the charge of François Bitouzet, who explains that the hyphenated name references its origins in Meursault on his mother's side (Prieur) and Volnay on his father's side. The vineyard holdings split along similar lines, with outposts to the north in Beaune and south in Puligny-Montrachet. An enviable line-up of *premiers crus* – three in Meursault and six in Volnay – dominate production, though the *village* Meursault and its Volnay sibling should not be overlooked.

'I make wines with my grapes from my plots, which gives me total control. I prefer to stay small to keep quality,' says Bitouzet, relating that he did once buy-in three barrels of Puligny-Montrachet 1er cru Champ Gains 2011: 'That was the first and last time – our own *village* wine was better.' For the white wines the juice is left to settle for 12 to 24 hours before racking to barrel, of which 20 per cent are new for the *village* wines, rising to 25 for the *premiers crus*. Barrels come from a quartet of suppliers: Seguin-Moreau, François Frères, Damy and Cavin, all standard 228-litre *pièces*, though Bitouzet is currently doing a study to compare them with the larger 350-litre barrels that have become an alternative standard in recent years.

For the reds he de-stems 100 per cent – 'I don't like whole bunch' – and the berries go to stainless steel tank for five or six days' maceration at 11°C before fermentation. He has abandoned wooden vats because of the difficulty of keeping them clean – converting one into a tasting table and another into his office desk: 'It's not IKEA, though I prefer to sit on a tractor than here.' He does two *pigeages* per day early in fermentation and perhaps some *remontage* towards the end.

The results range from a deliciously brisk Aligoté, which Bitouzet suggests as a perfect accompaniment to a *casse-croûte* (literally 'break the crust') or snack in the early morning during harvest, through to a sweet and succulent Volnay 1er cru Clos des Chênes that finishes con-centrated and firm. Between those two poles comes a Meursault 1er cru Santenots that Bitouzet describes as, 'a tannic white wine,' and a soft and subtle Volnay 1er cru Les Aussy – it's not a *monopole* though he is the only vigneron to label his wine thus. (Bitouzet's holding in the *lieu-dit* Les Aussy lies within the *premier cru climat* Le Ronceret.) There is much to like in these well-crafted wines.

Try this: Volnay 1er cru Taillepieds
Taillepieds has one of the steepest slopes in Volnay and the Bitouzet-Prieur vines sit at an altitude of 290 metres, with an exposure a little south of east. Whether the vineyard workers risk pruning their feet – *tailler les pieds* – because of the slope is open to conjecture. What is not is the quality of this gracious wine, the most 'classic' Volnay in an impressive range of *premiers crus*. Oodles of savoury red fruit, lifted by a spicy note, is the abiding impression, before the soft tannins kick in and carry the wine on to a clean, fresh finish. Balance is everything in this wine – it is no shrinking violet yet no weightlifter either.

Domaine Fabien Coche

5 rue de Mazeray, 21190 Meursault
www.fabiencoche.com; tel: +33 6 86 90 38 09

When I first bought these wines, some 30-plus years ago, the domaine was called Coche-Debord, which changed to Coche-Bizouard and then to its current incarnation as Domaine Fabien Coche. Back then the style was correct, nutty and uncompromising, softened only by long-haul ageing and then emerging splendidly taut yet ample with gorgeous truffled fruit flavours.

The style is more accessible in youth now but the purity of flavour remains a constant right across the range, which includes some seldom heard of and tiny Meursault names. Indeed, one could be forgiven for thinking that Coche specializes in vineyard names that would make for good dining table quiz questions, such as: Le Pré de Manche, a fragmented vineyard north of Meursault, one segment of which abuts Monthélie, and Les Gruyaches, a tiny triangle that looks like it has been lopped off the *premier cru* Les Charmes Dessous. Thus he shines a light on Meursaults that we seldom hear of and which, in his hands, deliver wines of compelling intensity. The Pré de Manche holding lies across three terraces and the first parcel was bought by his grandfather. The nose is quiet, all the excitement here is on the tingling, citrus-infused palate. The Gruyaches carries the same signature but in a slightly richer, more fleshed-out package.

Meursault is so inextricably linked with white wines that the reds are consistently ignored. Yet almost every scrap of Meursault land,

202 CÔTE D'OR

including celebrated *premiers crus* such as Perrières, Genevrières and Gouttes d'Or could be planted with Pinot Noir. Granted, the reds were once justifiably dismissed as rustic country cousins, but to adopt the same mindset today is to ignore some deliciously fresh, well-made wines. Coche's Meursault Rouge is just such a wine, immediately like-able and consistently satisfying thanks to a pleasant rasp of tannin and juicy red berry flavours that call out to be quaffed. The basic Bourgogne Côte d'Or Rouge, made from two parcels in Meursault and one in nearby Pommard, marches to the same beat with lovely crunchy fruit flavours. These wines may not be profound but they are worthy of attention nonetheless.

Grander wines are also made by Coche, such as the Meursault Goutte d'Or, which carries a creamy overlay of richness on the signature citrus of the house style before finishing with great length, but to ignore the entry level at this domaine would be a mistake.

Try this: Meursault Les Chevalières

Though Meursault possesses no *grands crus*, it lays claim, at the other end of the scale, to a panoply of *village* vineyards that leaves other communes in the shade. The 9-hectare Chevalières is just such a vineyard. It lies north-west of the village and the name may reference the fact that horses were once bred here. Fabien Coche produces a wine that punches at *premier cru* level, built around a core of mineral intensity, complemented and fleshed out by ripe fruit, while notes of spice and savour add complexity.

Maison Vincent Girardin

ZA Les Champs Lins BP 48, Impasse des Lamponnes, 21190 Meursault
www.vincentgirardin.com; tel: +33 3 80 20 81 00

Vincent Girardin started in business in a small way in his home village of Santenay in 1984. Quickly showing himself to be an astute businessman he expanded to the point where he needed larger premises, so moved to Meursault in 2001. A decade later, having established a sound reputation, particularly for his white wines, he caught everybody by surprise by selling up and departing Burgundy for agricultural enterprise elsewhere. Today, the business is part of Jean-Claude Boisset, though is largely autonomous (and should not be confused with Domaine Pierre-Vincent Girardin, Vincent's son's business).

THE VILLAGES AND PRODUCERS OF THE CÔTE DE BEAUNE 203

One constant at Girardin for almost all this century has been oenologist Eric Germain, who was appointed in 2002 and is determined to restore the reputation of white burgundy, as the dark shadow of premature oxidation recedes. 'The goal is to obtain wine for ageing', he says, explaining that the grapes are crushed before pressing to obtain a high level of lees to protect the wine. He is also holding back some production for later release: 'We need to come back to this way of consumption after premox, we should be drinking properly aged wine.' A taste of the Chassagne-Montrachet 1er cru La Romanée 2013, drunk in early 2024, endorses Germain's point elegantly. It was aged, yes, but in a positive way, sumptuous and mouthfilling, with a mild hint of truffles on the finish.

Paradoxically, where the Girardin white wines used to garner all the praise, it is now the reds that call for attention. This is not to suggest that the whites have declined in quality, if anything the opposite is true, but to acknowledge the great leap in quality achieved by the reds in recent years. They now display real character, no longer trailing in the shadow of the whites.

De-stemming 'depends on the vintage and the grapes' character', says Germain. No whole bunches were used in 2023, for instance, while 100 per cent were used in 2019. Vinification starts with four days' cool maceration in stainless steel, followed by fermentation with a combination of punching down early on and pumping over thereafter. Barrels are one-third new, the remainder being second and third fill. The results are seen to impressive effect in two Volnay *premiers crus*: Les Santenots and Les Champans. The Santenots vineyard lies within Meursault but if the wine produced there is red, it is labelled as Volnay. Leaving aside such confusion, this is a vigorously flavoured, well balanced wine that majors on intense red fruit leading into a pleasant finish. The Champans takes things up a notch, with grippy dark fruit, some exotic spice and firm tannin that needs a decade in the cellar to shine: the building blocks are all there. The message is clear – Girardin is no longer a one-colour pony.

Try this: Meursault 1er cru Les Perrières

It is generally agreed that Perrières is *primus inter pares* amongst the Meursault *premiers crus*, carrying just that bit more conviction and depth of flavour. The vineyard comprises four sub-sections – Aux Perrières, Perrières Dessus, Perrières Dessous and Clos des Perrières – and lies on

204 CÔTE D'OR

the southern boundary of Meursault, rubbing shoulders with Blagny. The Girardin Perrières starts mild on the nose, then develops slowly on the palate, succulent and plump at first with mild spice and a citrus ping chiming in to add balance. There is weight and substance, it is lush but not overbearing and it finishes with a long, rich flourish.

Domaine Patrick Javillier

9 bis rue des Forges, 21190 Meursault
www.patrickjavillier.com; tel: +33 3 80 21 27 87

'Our philosophy has not changed but we are adapting to conditions,' says Marion Javillier, daughter of the now retired Patrick. Elaborating on the challenges thrown up by climate change, she continues: 'Harvest is earlier ... We need to adapt, each year is different, there are more variations now from vintage to vintage ... The 2021 season was just awful, with really hard work in the vineyards.' It is a message echoed by her fellow winemakers the length of the Côte d'Or.

They do things a little differently at Javillier. While the trend in almost every domaine in the côte is for greater separation of parcels and the vinification of tiny plots – if they can be perceived to be marginally different from neighbouring ones – the opposite is true here. The house style derives from the blending of complementary plots that other domaines would keep separate. And it is not just the plots that are blended, their names are too, combined portmanteau fashion to yield results that sound like genuine *lieux-dits* but which will be searched for fruitlessly on the map. Witness the Meursault Les Clousots, which gets its name from a combination of Les Clous and Les Crotots; or the Meursault Tête de Murger, a 'cross' of Les Casse-Têtes and Au Murger de Monthélie.

The domaine is now managed by Marion and her brother-in-law Pierre-Emmanuel Lamy who is married to her sister, Laurène. It extends to 11 hectares covering 14 appellations, with production splitting 80:20 in favour of white wines. Sensitive use of oak is a domaine hallmark. 'Oak is a support,' says Marion, further explaining, 'if there is more clay in the vineyard we use less oak, more limestone and we use more wood'. In an unexpected inversion of more normal practice the top wine, a Corton-Charlemagne, sees no new oak at all, new barrels having their precocious youth softened by being seasoned with the *regional* and *village* wines before moving up the scale.

The wines are more polished than precise and display the house style of smooth-flavoured elegance, rather than sharp, terroir-induced differences, thanks to the blending of different *lieux-dits*. Save, that is, for the Corton-Charlemagne, which strides across the palate rich and generous, with ample fruit and good depth. The remainder of the range may be sound rather than special but you are in safe hands with Javillier.

Try this: Meursault Les Tillets

This 12-hectare vineyard sits at the top of the slope, east facing and overlooking the *premiers crus*, and reaches 350 metres elevation, where it abuts the treeline. Javillier is the largest owner, with more than 1.5 hectares, which represents the domaine's biggest vineyard holding. Compared to Javillier's citrus-infused Clos du Cromin, Les Tillets is fuller in every dimension and, although there is a firm mineral note, it leans more towards what might be termed the 'butter and nuts' flavour profile of classic Meursault. But it only leans, this is still a racy, energetic wine.

Domaine des Comtes Lafon

Clos de la Barre, 5 rue Pierre Joigneaux, 21190 Meursault

www.comtes-lafon.fr; tel: +33 3 80 21 22 17

Jules Joseph Barthélémy Lafon was born in south-west France in 1864, and 30 years later married Marie Boch, whose family were winemakers in Meursault. Over the next 40 years he gradually built up the domaine holdings by way of astute purchases of prime vineyards, including a roughly square plot of 0.32 hectares in the south-east corner of Montrachet, laying the foundations of today's celebrated domaine. Jules' great-grandson, Dominique Lafon, was in charge here from 1985 before retiring in 2022 and handing over to his daughter, Léa, and nephew, Pierre, though he remains a background presence for advice and consultation.

Pierre worked previously in the United States before returning to Meursault in 2020, to what he readily admits was a steep learning curve: 'I had been selling wine in Chicago but I learned so much more here after I returned.' Candour such as this imbues all his conversation, frequent references emphasizing his open-minded approach to issues such as the use of whole bunches for the red wine fermentation: 'We used some whole bunches in 2020, 50 per cent in 2022 and de-stemmed 100 per cent in 2023. We are flexible.'

206 CÔTE D'OR

Regardless of the de-stemming policy the Lafon reds have a polish, a gloss in the texture that endows immediate appeal, even when tasted from barrel. The Pommard Les Vaumuriens comes from a high and cool vineyard, well above the village, the sort of site that is benefiting from climate change. Made for the first time by Lafon in 2023 it also benefits from a sensitive approach: 'We wanted to avoid making a big red, so used a light extraction. Taking on a new plot is a learning process.' Judging from the bountiful fruit and vibrant finish the Lafons are quick learners.

Understandably, the Lafon white wines still garner most of the attention but that is due to the appellations being of a higher rank, not winemaking dexterity. The vineyards have been farmed organically since 1995 and biodynamically since 1998. Once picked, the Chardonnay grapes go straight to the press in whole bunches and a lot of the lees are retained. Fermentation begins in tank before transfer to barrel, principally 228-litre *pièces*, 20–30 per cent new, though an increasing number of 500-litre *demi-muids* are to be found in the cellar: 'It's a great container', says Pierre.

'Tension' is a word Lafon uses frequently to describe the style of the whites, a quality seen to good effect in the Meursault 1er cru Les Porusots, defined as it is by lively acidity and an almost tannic dryness on the finish. Less austere, though still displaying great energy, is the Puligny-Montrachet Les Charmes, which first came into the Lafon fold in 2022, courtesy of a long-term *fermage* contract. Numerous other *premiers crus* of increasing grandeur lead ultimately to the seamless splendour of the Montrachet: powerful and poised, rich and full, ample in every dimension, most particularly the ringing finish.

In the persons of Léa and Pierre Lafon the domaine's legacy is secure. Might the new generation find time to revive the domaine's moribund website? 'That is something we are supposed to do. Maybe next year …'.

Try this: Volnay 1er cru Santenots du Milieu

Lafon is renowned as a white wine producer, yet the domaine's Volnay Santenots holding of 3.8 hectares is its largest by some margin. The wine is one of the Côte de Beaune's best reds, rich and deep, full and firm, but not weighty, thanks to the incisive acidity. There's robust tannin too, and concentrated fruit. Yet despite the abundant components there is no lack of finesse. A good vintage should not be broached before 10 years of age and can last much longer. Look also for the 'junior brother', made from young vines planted in 2002 and 2015 and declassified to *village* to deliver a lighter iteration, a delicious outline sketch rather than the full painting.

Domaine Vincent Latour

6 rue du 8 Mai 1945, 21190 Meursault

www.domaine-vincentlatour.com; tel: +33 3 80 21 22 49

Visitors here will do a double take, wondering if they have landed in a Chartreuse cellar rather than a wine cellar. An impressive collection of bottles of the noble spirit takes centre stage and Vincent Latour will happily expound at length about them. It can take a little effort to steer the conversation back to wine.

The family traces its origins in Meursault back to Revolutionary times, though the domaine as currently constituted began to take shape in 1998. Today, Vincent and his wife, Cécile, have charge of 9 hectares of vineyards, principally located in Meursault but also Pommard, Saint-Aubin and Chassagne-Montrachet, with their sons, Thomas and Nathan, waiting in the wings.

Vincent Latour likes to say that from 1998 to 2004 he was finding his way and that since then he knows the way he wants to go – and that way does not include much new oak. The white wines see about a year in wood, very little new, including 600-litre *demi-muids*, with no bâtonnage, followed by four to six months in stainless steel tank before bottling. A pair of named cuvées are flag carriers for the succulent yet refined house style: Bourgogne Côte d'Or Cuvée Héritage and Saint-Aubin Cuvée Thomas. The Bourgogne is made from vines planted by Vincent's grandfather, in a pair of vineyards in 1951 and 1956. The texture is suave and the fruit is ripe, with a touch of mild spice to add lasting interest. The Saint-Aubin, named for '*mon premier fils*', comes from the centre of the commune and oscillates beautifully between sweetness and savour, with a dart of minerality as ringmaster. Further up the scale, the Meursault 1er cru Les Perrières, from a 0.17-hectare holding next to Puligny-Montrachet, is treated to 50 per cent new oak – unusually high for Latour – to deliver a lavishly-flavoured wine that would be overly lush were it not for the sweet lemon at its core.

Latour also produces some attractive red wines, notably a Volnay Les Serpens and a Meursault 1er cru Les Cras. Both are made with 100 per cent de-stemmed grapes. The first has tangy acidity cloaked in glossy fruit, while the second has a firmer texture without being in any way

208   CÔTE D'OR

coarse. The domaine's reputation rests on the quality of its whites, however, and it could be said that they play a potentially dangerous game, as their flavours lean towards lush opulence that could easily tumble into cloying richness. They avoid that trap thanks to the discreet tingle of acidity that each has at its core.

Try this: Meursault 1er cru Les Charmes

At 31 hectares Charmes is the largest of the Meursault *premiers crus* and Latour owns one-third of a hectare, divided between the lower (*dessous*) and upper (*dessus*) sections. Charmes is noted for wines of immediate appeal, which in this case sees it fitting seamlessly with the Latour house style. Production runs to about one foudre and four barrels and there is very little new oak used, if any, allowing the lush fruit to shine. The impressive depth of juicy fruit leans towards tropical, with a whiff of Turkish delight opulence, all given focus by a mouth-watering trickle of acidity.

Domaine Latour-Giraud

6 rue de l'Hôpital, 21190 Meursault

www.domaine-latour-giraud.com; tel: +33 3 80 21 21 43

Jean-Pierre Latour shrugs: 'We are very tired, both physically and psychologically, it has been tough.' He is speaking during the 2024 growing season, when rampant mildew challenged the vignerons almost like never before, some saying it was the worst they had seen since 1993, which, coincidentally, was the year Latour took over from his father here.

On a brighter note, his switch to organic viticulture, gaining certification in 2024, has been an unqualified success. The wines were always consistent and easy to like but now they display greater precision and accuracy, they are more compelling, with vibrancy and excitement added to the flavour palette. Latour started to employ organic practices in 2018 but shied away from certification because of the paperwork: 'I did it quietly to check if it was an advantage or not. Now that I see it, I am prepared to take on the admin side,' he says, with an eye-roll, elaborating: 'It is a new step for us to take … It gives us a clear picture of the terroir … At the end of élevage I have seen greater character and fruit flavours … There is more precision and more energy in the wine.' These are not trite observations, crafted to serve as sales and marketing blather, Latour's wines give telling confirmation to his words.

Eighty-five per cent of production is white, the grapes going straight to the press, with no crushing, after which the juice is allowed to settle for two days in stainless steel, cooled initially to 16°C and then allowed to start fermentation in tank before transfer to standard, 228-litre Burgundy *pièces*. Latour does not want any overt oak influence so uses about 20 per cent new barrels for the basic wines, increasing to 25 per cent for the *premiers crus*. He would like to use 350-litre barrels but lack of space precludes that.

The domaine is the second-largest owner of Meursault 1er cru Les Genevrières with 2.5 hectares. Latour explains the challenge of harnessing its potential: 'It is a difficult terroir because you can get a lot of energy but lose precision, it doesn't support too much oak. It is difficult to make if you want to retain all the elements that make this wine concentrated yet fragile.' He turns a deft hand to the challenge, resulting in a multi-faceted wine that delivers a wave of flavour nuances all the way into the lingering finish.

Latour's red wines should not be ignored, especially the Volnay 1er cru Clos des Chênes: 'It is one of the stronger *premiers crus* of Volnay, it can age a long time and keep all the elements we expect from Volnay.' There is perky fruit, juicy acid and a pleasant rasp of tannin to corroborate those words.

Jean-Pierre Latour's daughter, Josephine, is now working alongside her father, though he is in no doubt about the challenges that face the next generation: 'I want her to understand the wine side and then work on the business part. The complete work of a winemaker is too much for one person today, my wife does all the administration. When we retire Josephine will need to employ someone to do the admin.' Wise words from a man whose wines don't get quite the recognition they deserve.

Try this: Meursault 1er cru Les Bouchères
Bouchères might be regarded as the hidden *premier cru* of Meursault, probably because of its diminutive size, about 4.5 hectares compared to the 30-plus hectare Charmes. Jean-Pierre Latour emphasizes the importance of getting the picking date correct: 'It can be heavy if you get the picking window wrong, two or three days late and you lose the style of the origin.' That style is fine and pure, without the amplitude sometimes associated with Meursault, but with a discreet creamy texture to add elegance and complexity.

Domaine Matrot

12 rue de Martray, 21190 Meursault
www.matrot.com; tel: +33 3 80 21 20 13

'We want to make classic cuvées with round texture and minerality to keep the wine fresh and straight,' says Elsa Matrot who, along with her sister, Adèle, represents the sixth generation at this family domaine. Elsa joined her parents, Thierry and Pascale, here in 2008, followed by Adèle two years later. Elsa had no vinous ambitions when growing up and it was only after she started to study law and found it not to her liking that she changed direction and set about learning the business, from planting the vines to selling the wines: 'I wanted to make wine before selling it, I couldn't talk about it unless I was making it.'

Production from the domaine's two dozen hectares splits 60:40 in favour of white wines, from vineyards largely situated in Meursault, Puligny-Montrachet and Auxey-Duresses. Cultivation has been organic since 2000 and the domaine is due to achieve certification in 2025. Elsa is acutely aware of the challenges posed by climate change and the vines are being pruned later than previously in an effort to lessen the threat of spring frost, 'but we can't prune 24 hectares in March'. Similarly, replanting decisions call for great consideration given that they last 'for decades'. Elsa's thinking is that it might make sense to replant at 9,000 or even 7,000 vines per hectare, because of drought conditions that affect young vines disproportionately, a mindset that goes against the recent trend to increase planting density.

The Matrot wines reflect this cerebral approach, from the Bourgogne Chardonnay up to the patrician Puligny-Montrachet 1er cru Les Chalumeaux. The Bourgogne sees more new oak, about 20 per cent, than the other wines, to season the barrels, and is a 'tripartite' wine whose components come principally from Meursault and the Hautes-Côtes de Beaune with, somewhat surprisingly, a small component of must from Chablis (and not Mâcon, Matrot adds because, 'it is too heavy'). The Meursault 1er cru Les Charmes is an essay in reserved opulence; ample fruit balanced by a saline savour sets up a delicious sweet and sour interplay, with the lucky palate as the beneficiary. By comparison, the Puligny is svelte and smooth, polished in every respect but perhaps a little less engaging – a preference for one or the other will be entirely personal.

The reds may lack the grace of the whites but they are attractive, despite a less complex flavour register. They are all made with 100 per cent de-stemmed fruit and the leading wine is the Blagny (below) though the Auxey-Duresses should not be ignored: 'It can be a little rustic and tough. We wait for maturity to have softer tannins,' says Elsa Matrot. Climate change has helped this higher altitude vineyard achieve better ripeness: 'There is much more roundness in the wine than before ... We want elegant and straight, something very fine, not transparent but not over-extracted.'

Matrot concludes: 'We like acidity, we like crispy wine but it must be balanced ... We do a classic vinification, we follow the style of our father.' Yes, though there is no doubt that the wines have moved up a notch in recent years, courtesy of greater elegance.

Try this: Blagny 1er cru La Pièce Sous le Bois

Blagny is a tiny hamlet that lies above the village of Puligny-Montrachet and on the boundary with Meursault. Though the name suggests that this vineyard lies directly below the wood, it is separated from it by a tiny strip of another vineyard, Le Bois de Blagny. 'It is our flagship red,' says Elsa Matrot, whose great-grandfather, Claude, bought the domaine's holding in 1900. Some Blagnys can be a bit sinewy but not this one, whose core of lively acidity is well wrapped in rich and round dark fruit, with firm tannins to endow longevity.

Château de Meursault

5 rue du Moulin Foulot, 21190 Meursault

www.chateau-meursault.com; tel: +33 3 80 26 22 75

Château de Meursault's history stretches back to the eleventh century, and the extensive cellars, which can accommodate 800,000 bottles and 2,000 barrels of wine, date from the twelfth, fourteenth and sixteenth centuries. In this case the term 'château' is apt, for this is one of the few Côte d'Or estates with grand buildings and an expanse of vineyards on a par with those in Bordeaux. And the château has got a whole lot grander since the first edition of this book.

After occupation by German troops during the Second World War it was left empty and fell into an increasing state of disrepair until a complete renovation programme was started in 2022. The entire interior was gutted, including the removal of all the floors, to leave just a shell that was then brought back to sparkling life, housing meeting rooms,

212 CÔTE D'OR

tasting rooms, a professional kitchen and comprehensive visitor facilities. The visitor experience, previously satisfactory, has been elevated to one of the best in the Côte d'Or. More pertinent to the winemaking, l'Ancienne Cuverie, better known to many as La Salle de la Paulée, venue for the annual Paulée de Meursault, is being converted back into a working winery, to be used for the *premiers crus* and *grands crus* red wines. (Regular attendees at the paulée needn't worry, it will still be able to host the bibulous celebration.)

With harvests now happening earlier but, more pertinently, faster than previously it had become imperative to process the grapes more quickly. With the new facilities the entire harvest should be processed in eight to ten days, down from about twelve previously. The expansion was also necessitated by an increase in the estate's holdings, up to 67 hectares from 60, spread across 100 plots. It would be difficult to overstate the extent of developments at Château de Meursault in recent years, driven and overseen by Stéphane Follin-Arbelet, appointed as manager by Olivier Halley shortly after he acquired the property in 2012. The Clos du Château – with only a little licence it can be called the front garden – is now a showcase for the biodiversity that Follin-Arbelet is seeking to establish. To that end, 3,000 plants, shrubs and trees have been planted there, creating a horticulturalist's paradise, a detailed description of which lies beyond the scope of this book.

Almost immediately after Follin-Arbelet's appointment commentators noted an uptick in the quality of the white wines, while agreeing that the reds needed a similar lift to match their siblings. This has now happened, largely thanks to a more sensitive approach to vinification. After harvest by hand into 15-kilogram boxes the grapes are usually de-stemmed before fermentation in large wooden vats, with *remontage* and *pigeage* use indicated by the quality of the vintage. 'We don't have a recipe,' says Follin-Arbelet, 'we want to avoid a big extraction, we want the wines to be delicious from beginning to end of the drinking window. We don't filter them so as to keep the silky tannins.'

In that respect the Beaune-Grèves 1er cru Les Trois Journaux is a herald for the new, more gracious style, replete as it is with abundant red berry fruit supported by civil tannins and given focus by a gentle spine of acidity. A huge amount has been achieved at Château de Meursault, yet rather than feeling like a job well done, it points the way forward to similarly ambitious efforts in the future.

THE VILLAGES AND PRODUCERS OF THE CÔTE DE BEAUNE 213

Try this: Meursault 1er cru Les Charmes-Dessus

Two parcels, amounting to a generous total of three hectares, make up the château's holding in this, the favoured upper section of the Charmes vineyard. The slope has a steeper gradient than the lower section and faces almost directly east. Waxy fruit on the nose presages a palate that manages the sometimes difficult marriage between elegance and substance to yield an intensely flavoured mouthful. Rich fruit is kept lively by a mild touch of spice that carries the wine clean and intense all the way to the long finish. 'The only weakness is that you want to drink the whole bottle,' says Stéphane Follin-Arbelet.

Domaine de Montille

8 rue du Cromin, 21190 Meursault

www.demontille.com; tel: +33 3 80 21 39 14

Tracing de Montille's history back to the mid-nineteenth century reveals a domaine of impressive size, whose vineyards included holdings in Côte de Nuits *grands crus* such as Musigny and Bonnes Mares, as well as a significant footprint in the Côte de Beaune. A century later, post-phylloxera and the other tribulations that plagued the wine business, such as world wars and economic depression, family fortunes had taken a dramatic turn for the worse. The domaine had shrunk to 3 hectares, mainly in Volnay, leaving Hubert de Montille with the dual challenge of tending to the vines while also pursuing a career in the law. He did both successfully and thanks to his efforts and those of his son, Etienne, grandeur has returned to Domaine de Montille, by way of vineyard holdings that total 37 hectares and cover nearly three dozen appellations.

Today, de Montille is a tri-located domaine with bases of operations in Volnay, Meursault and Puligny-Montrachet – as witness the bottling machine mounted on a wheeled jig to allow easy transport to where it is needed. The scale of operations makes it one of the largest family domaines in the Côte d'Or – a scale that was given a considerable boost in 2012, when Etienne de Montille bought the Château de Puligny-Montrachet, having managed it for the previous decade. There is also a small négociant business, Maison de Montille, formerly known as Deux Montille when Etienne's sister, Alix, was involved. This name change ends the confusion that arose when saying 'de Montille' and 'Deux Montille', for there was only a wafer of verbal separation between them.

214 CÔTE D'OR

Saying 'dux', rather than holding up two fingers, was the easiest way to clarify things.

In charge of day-to-day operations is the chatty and wisecracking American Brian Sieve, who followed a circuitous career route before starting here in 2010. Production splits 50:50 between white and red wines, purity and finesse being their twin hallmarks. One of the leading white wines is also, on paper, one of the humblest: the Bourgogne Blanc, Le Clos du Château. Produced from a 5-hectare vineyard that used to be an orchard, and situated in front of the Château de Puligny-Montrachet, this walled vineyard produces a wine of immediate, fruity appeal with enough mineral savour to see it easily mistaken for a Puligny *village*. At the top of the white tree sits the Corton-Charlemagne, the 1-hectare plot of which was bought in 2004. It was originally planted with Pinot Noir but this was changed to Chardonnay by de Montille. Judging by the composed and elegant palate, with a saline seam running through it, the decision was a wise one.

Despite the quality of the white wines the reds still carry the flag for de Montille. A seldom-seen rarity is the Beaune 1er cru Les Sizies, an 8.5-hectare vineyard that sits roughly in the middle of the Beaune *premiers crus*. Apart from the expected cherry fruit there is a pleasant perk of acid and a mild cut of tannin, all of which resolve into a clean and fresh finish. Meanwhile, the one hectare of Volnay 1er cru Les Champans produces a wine that could serve as a template for everything that good Volnay should be. Scented fruit carries into a vigorous but balanced palate replete with cherries and raspberries, a mild note of spice, all borne along by a discreet tannic foundation. These are but two of a wide and impressive range that rewards repeated investigation.

Try this: Pommard 1er Cru Les Pézerolles

The name Pézerolles derives from '*poizerolles*', meaning chickpea in ancient French, which suggests that this was the crop here before vines were planted. About one-quarter of the de Montille vines are 70 years old, a further half are 30 to 40 years and the remainder were planted a dozen years ago. De-stemming is dictated by vine age, giving an average of one-third whole bunch used. The supple and elegant style of this wine makes it something of an outsider in the traditional Pommard firmament. Taste alone would suggest that it lies close to Volnay but the opposite is true, for it is almost on the boundary with Beaune. There is ample dark fruit, smooth tannins and no heft at all.

Domaine Roulot

1 rue Charles Giraud, 21190 Meursault

Tel: +33 3 80 21 21 65

Former actor Jean-Marc Roulot was gifted his 15 minutes of YouTube fame during the Covid pandemic when Donald Trump's advocacy of disinfectant to fight the virus handed him a too-good-to-miss opportunity to call out Trump by way of a searing impersonation. Amusing though this was, one should be in no doubt that Roulot is a winemaker on top of his game.

He took over the reins of this family domaine in 1989 and in the decades since has secured for it a place in the pantheon of white burgundy producers. Extraordinary purity and definition of flavour is the Roulot hallmark, expressed in a range of wines that stretches from a super-intense Bourgogne Blanc (below) through a clutch of *village* and *premiers crus*, finishing with the almost impossibly flavoursome Les Perrières. When tasting these wines it is difficult to grasp the fact that the family once made their money from distillation and it was those earnings that helped to expand the vineyard holdings.

Roulot was joined by his son, Félicien, in 2022 and, if such is possible, the wines seem more assured than ever today. The grapes are crushed before pressing, with the juice allowed to settle overnight in stainless steel before transfer to barrel for fermentation. A year later, before the next harvest, the wine is racked back into tank for six to seven months prior to bottling. Thus the barrels, perhaps no more than 20 per cent new, always remain in use, and are considered to be at their best when one year old. A small number of amphorae and porcelain barrels are also used. Roulot's assistant, Paul Delorme explains: 'The wine is not as complex from amphorae, we use it like a spice; the porcelain gives good energy.'

If one word could be used to describe the style it is 'energy'. The wines are lean but not austere, clearly delineated, upright and reserved, yet abundantly flavoured; they possess a nobility that sets them apart. A trio of Meursault *premiers crus* exemplify that template: Porusot, Clos des Bouchères and Les Perrières. The Porusot comes from two plots and delivers a palate of citrus purity with an added creamy note to soften the edges; the Bouchères *monopole* is more complex with a marvellous textural polish to the seamless palate

216   CÔTE D'OR

profile; the Perrières is complete and generous, carrying some added flesh but not a hint of fat.

With age, the wines unfold and open, the linear precision broadening and developing extra dimensions – fatness and succulence, elements of truffled fruit – while somehow encompassing the contrasting characteristics of leanness and richness. With that in mind a new cellar is due for completion in 2025 to allow for quantities of the top wines to be held back for some years before release, answering a concern of Roulot's that if his wines are drunk at around four to five years old they will be closed and not showing at their best. It will also rationalize a set of buildings for which the term 'higgledy-piggledy' could have been coined.

Try this: Bourgogne Blanc

When selecting a single wine to recommend from the Roulot stable one is confronted by a surfeit of riches, yet for a simple test of a domaine's wines it is often best to check out the basic, entry level offering. This Bourgogne Blanc is a wine of remarkable polish and finesse, a beautifully composed wine of seamless elegance and impressive length that accounts for one-third of the domaine's production. Perhaps the best acid test of its quality is to see how other winemakers' eyes light up when I serve it to them in a social setting.

PULIGNY-MONTRACHET

Puligny-Montrachet stands, *primus inter pares*, as the source of the greatest white burgundy, notwithstanding the compelling claims of Chassagne-Montrachet and Meursault to be considered as such. Puligny manages to ease its nose in front by way of a tad more grace, a morsel more conviction, a scintilla of greater satisfaction. This, it must be noted, is a general statement, individual producers can turn its truth on its head. But, when on song, there is refinement and restraint, opulence held in check by civilized acidity, sumptuous flavours that never obtrude and succulent fruit counterpoised by mild minerality. A great Puligny has it all.

The village is known the world over for white wines of peerless quality with a seamless character that weaves together the components into something greater than their sum. When properly aged – and not displaying the drab flavours of premature oxidation – the wines develop a

THE VILLAGES AND PRODUCERS OF THE CÔTE DE BEAUNE 217

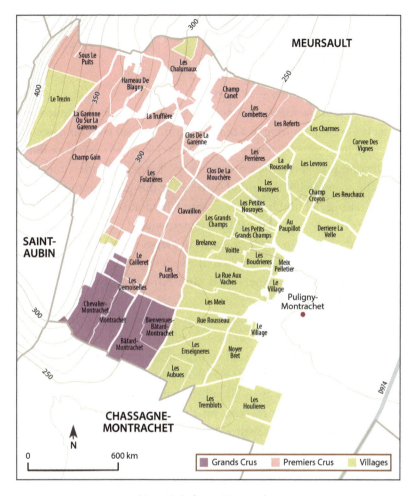

Map 11: Puligny-Montrachet

truffley succulence, with exotic scents and flavours that have the better of any superlative coined to capture them.

Puligny-Montrachet presents a prosperous face to the world, given emphasis in recent years by substantial improvements in the public spaces, particularly the Place des Marronniers, which has been partly pedestrianized, with flowerbeds and ornamental pool. Properties are in good order – the ramshackle, yet to be restored buildings that still dot some other villages are not seen here. And it is but an easy walk up a gentle slope to the Montrachet vineyard and its satellites, a 'bucket list' place of pilgrimage for visitors from across the globe.

218 CÔTE D'OR

Producers

Domaine François Carillon

2 place de l'Eglise, 21190 Puligny-Montrachet
www.francoiscarillon.com; tel: +33 3 80 21 00 80

Expect the unexpected at this domaine. 'What's that?' I enquired of François Carillon pointing to a bottle of rum, labelled with his name, as we settled in to taste his wines. 'A good experience,' he replied, smiling. Equally unexpected was the score of empty Chartreuse bottles lining the cellar and, completing the trio of surprises at this patrician domaine of ancient lineage, a humble Vin de France Chardonnay. This latter wine might be called a child of the 2021 vintage, when Carillon, having lost 80 per cent of his crop, went in search of grapes to pad out his meagre production of Puligny-Montrachet. The losses were cruel: one barrel of Les Perrières *premier cru* where 15 is more normal, 27 barrels of *village* when he might expect 120.

The Vin de France was made from grapes sourced near Nîmes, far to the south of Burgundy, and was labelled 'Cap au Sud' (Heading South). The grapes were pressed *in situ* and transported by tanker at 12°C for vinification in Puligny and so popular did the wine prove that Carillon has made it in all subsequent vintages. Out of adversity has come enduring success, yet the erratic challenges posed by climate change – notably spring frost and summer hail – call for adaptability and great fortitude, while the less erratic but no less taxing spectre of searing summers, leading to ever earlier harvests, prompts a sanguine response from Carillon: 'My grandfather never harvested in August, my father picked twice, 2003 and 2007, and I have stopped counting.'

Despite all the challenges, François Carillon, who worked for years alongside his brother Jacques (see next entry) before they went their separate ways in 2010, manages to produce an assured range of wines. The process is uncomplicated: after early morning hand harvest the grapes are pressed and the juice cooled to 12°C, settled in tank and transferred to barrels of varying sizes – 228, 350 and 450 litres, from *tonneliers* Chassin, Cavin and Claude Gillet – for about 10 months, then put back into tank for nine months before bottling.

All the wines have a svelte fruit countenance but if that was it they could be dismissed as one-dimensional with only fleeting attraction, which is far from the case. The Puligny-Montrachet *premier cru* Champ

THE VILLAGES AND PRODUCERS OF THE CÔTE DE BEAUNE 219

Gain sits at 350 metres elevation on poor, stony ground that brings superb intensity and a savoury bite to the house style. Meanwhile the deeper soil of Les Combettes yields a fuller wine, with a more succulent note on the palate.

In recent years François has been joined by his sons Mathis and Paul and, with retirement on the horizon, he quips, 'I look forward to watching them working,' before adding, 'I am still learning, so they have lots to learn as well.' And the rum? It is from Guadeloupe and is matured in Carillon barrels for two years – 'Finish Chassagne-Montrachet' – affording François a reason to travel there once a year.

Try this: Puligny-Montrachet Clos du Vieux Château

Thanks to the fact that this small *monopole* of three-quarters of a hectare is surrounded by three-metre high walls the grapes ripen quickly as harvest approaches, encouraged by the heat reflected off the stone. It was once a garden whose produce of cherries, apples and pears was reserved for the local lord and it lies little more than 50 metres behind the Carillon winery. Citrus and tropical fruits intermingle, while a lovely mineral bite holds the natural opulence in check right through to the fresh finish.

Domaine Jacques Carillon

1 impasse Drouhin, 21190 Puligny-Montrachet
www.jacques-carillon.com; tel: +33 3 80 21 01 30

Until New Year's Day 2010 Domaine Louis Carillon was an established part of the Puligny hierarchy; then brothers Jacques and François Carillon divided it between them, each striking out on his own, in premises barely 100 metres apart. Such division is a regular occurrence in the Côte d'Or; what made this unusual was that it happened many years after both had started to work with their father Louis (Jacques in 1980 and François in 1988) and not, as is more usually the case, when they were in their twenties.

Today, Jacques makes a modest seven different wines, all white, although he does grow some Pinot Noir, which he sells off as fruit. He did have a Saint-Aubin Rouge but has replaced those vines with Chardonnay. There is nothing flashy about him or his wines and, apart from picking a little later than others, there are no surprises in the vinification. After hand harvest the grapes go straight to the press, with the juice allowed to settle overnight in stainless steel at 8–10°C. The clear

220 CÔTE D'OR

juice then goes into standard *pièces*, 20 per cent new, for alcoholic and malolactic fermentation for a 'short' year, being racked back into tank three weeks before the next harvest. Some lees are kept at this stage before clarifying in January and bottling in February/March. Carillon is currently trialling DIAM closures and uses them on his Puligny-Montrachet *village* if an importer requests so. He will continue trials but until he is satisfied that the closure is good for 10 to 15 years cellarage he will stick to natural cork.

The Puligny accounts for a major share of production. It is a wine of harmony and elegance that exhibits the fine but pronounced character that marks all of Jacques Carillon's wines. It comes from a scattering of seven *lieux-dits*, which together amount to 2.7 hectares of vines approaching their half-century and, while always cleaving to the house style, it displays clear vintage variations, such as between 2021 and 2022. The first was born from difficulty, Carillon describing the dismal growing season as, 'a classic year for Burgundy in the 1980s,' and the wine carries a lean note, while the second, from an excellent vintage, is fuller and more ample.

At a more exalted level the Bienvenues-Bâtard-Montrachet, made from a 0.11-hectare holding that forms a narrow strip running from top to bottom in the vineyard, is surprisingly gentle on the nose, saving all the excitement for the palate, where the concentrated flavour pyrotechnics run deep rather than broad.

If Carillon's daughter, Alice, who 'helps out when needed' joins him full-time at the domaine then the future is assured and there are unlikely to be any major changes to his philosophy of 'try to do better each year'. He is concerned, however, that it is getting increasingly difficult to recruit workers willing to engage with the necessarily repetitive grind required to make wine at this level.

Try this: Puligny-Montrachet 1er cru Les Perrières

'I don't make my wines to drink young, I make them to age,' says Jacques Carillon as we compare the relative merits of this wine and the next-door *premier cru* Les Referts. The latter seemed buttery when compared to the tingling intensity of the Perrières, which is both rich and mineral in equal parts with lovely interplay between the two. In youth, that is. Approaching a decade old the 2014 bore out Carillon's words. It was still fresh, showing maturity rather than age, with some pleasant nutty notes leavening the fruit and adding complexity on the finish.

Domaine Jean Chartron

8 bis Grande Rue, 21190 Puligny-Montrachet
www.bourgogne-chartron.com; tel: +33 3 80 21 99 19

Jean-Edouard Dupard, a cooper, is remembered today for two things: founding this domaine in 1859, and successfully petitioning to have the Montrachet name added to Puligny in 1873, when he was mayor of the village. His daughter married a Chartron and today, five generations later, the domaine is in the hands of Jean-Michel Chartron and his sister Anne-Laure. He is a familiar figure in Puligny and played a prominent role in the organization of the Saint Vincent Tournante 2021 there, postponed to 2022 because of the Covid pandemic.

As with all Chartron's contemporaries he is much concerned with the damaging effects of climate change but while many talk in general, sometimes philosophical, terms about the challenges it poses he lays bare the nitty-gritty details. For instance, it costs €70,000 per annum to insure the 14-hectare domaine against all climate risks such as frost, hail and flooding (it is cheaper to insure the whole domaine rather than trying to identify the most at-risk sections). He is also exploring the possibility of using individual plastic canopies (trade name 'Wine Protect') for each vine that cost about €10 each, and which can protect against frost down to -6°C. They must be applied one-by-one and his greatest concern is the care needed not to damage the vine when they are being removed.

Such considerations aside, Chartron can count himself lucky that his roster of vineyard holdings contains more than the usual sprinkling of *grands* and *premiers crus* in amongst the humbler *village* appellations. The Clos des Chevaliers (below) is the flagship but there is also a pair of other *grands crus* – Bâtard-Montrachet and Corton-Charlemagne – and a trio of Puligny *premiers*: Clos du Cailleret, Les Folatières and Clos de la Pucelle. In addition, Chartron does some exchanges with other growers, for instance in 2023 trading eight barrels from the three *premiers* – Cailleret (3), Folatières (3) and Pucelle (2) – for one barrel of Montrachet.

The winemaking is straightforward: picking usually starts at 6.30 a.m. and finishes at noon, the whole bunches going straight to the press and the juice settled at 12°C overnight in tank. Standard 228-litre barrels are used for fermentation, up to 30 per cent new for the *grands crus*, and after about one year the wines are pumped back into tank, 'to make

222 CÔTE D'OR

room for the next harvest,' says Chartron. He goes on to explain that, being in Puligny with its high water table, he cannot work by gravity so must use a peristaltic pump and, 'every time I use a pump I wait for a month for the wine to recover'.

The results comprise a range of wines that satisfy rather than dazzle, with greater amplitude and conviction coming as you move up through the range. Included in their number is a rare red Puligny-Montrachet 1er cru Clos du Cailleret, made from a garden-sized plot of Pinot vines in the otherwise Chardonnay vineyard. It's elegant, if hardly compelling, and slides across the palate gently without much penetration. It is also possible to buy the Chartron wines at cellar door, something of a rarity in the Côte d'Or.

Try this: Chevalier-Montrachet Clos des Chevaliers grand cru

Jean-Michel Chartron produces about six barrels of this wine from his half-hectare holding, easily identified thanks to its arched stone entrance. 'This is the jewel in the crown, the parcel we are most proud of,' he says as he pours a sample. The wine plays to type and, as might be expected, exhibits more restraint than the Bâtard-Montrachet that fairly bounds across the palate by comparison with the Chevalier's more measured stride. A little severe and intense at first, the palate gradually opens to reveal tingling fruit and great length. It's racy and precise, chiselled and flinty, and only with age does it gain fullness to complement the firmness.

Domaine Leflaive

7 rue de l'église, 21190 Puligny-Montrachet
www.leflaive.fr; tel: +33 3 80 21 30 13

It is hard to avoid a repetitive caveat when writing of Domaine Leflaive, for this was one of the producers most seriously affected by the scourge of premature oxidation. Too many otherwise laudatory comments and observations about post-1995 vintages are qualified by the likes of: '… assuming the bottle is not oxidized.' Premox is the elephant in the room in any conversation with Brice de la Morandière, who succeeded his cousin Anne-Claude Leflaive at the helm here, after her early death in 2015. It is an elephant he has acknowledged without hesitation, and he deserves credit for eschewing a head-in-the-sand response to the problem. Head-in-sand was never going to solve it.

Morandière readily admits that he had no idea about premox 10 years ago but it took no more than a week 'to understand we had a big problem'. An immediate switch to DIAM closures, and the appointment of Pierre Vincent (see p. 195) as winemaker in 2017, were the changes most obvious to consumers but behind the scenes every aspect of the winemaking process, 'from hand on vine to hand on bottle', was subject to scrutiny. 'But I don't want to brag, we have to be humble,' Morandière concludes, before moving on to a tasting of the wines.

Some of the Leflaive *premier* and *grand cru* holdings are eye-watering. Rounding up a little: 5 hectares of Clavoillon, 3 of Les Pucelles and 2 each in Bâtard- and Chevalier-Montrachet; the 0.08-hectare postage stamp in Montrachet is more modest and yields about two barrels in a good year.

Tasting across the range is always a treat, a masterclass in terroir differences, given a house signature by way of exemplary balance and suave texture. The Bourgogne Blanc is pure and poised and it is only by comparison with its 'elders' that it might be described as a little one-dimensional. The Puligny *village* is clean and fresh with some mild richness adding a little weight. Amongst the *premiers crus* the Clavoillon is fuller, with lovely texture to embellish the singing fruit and mild spice, while Les Combettes moves things up a notch with greater depth and length adding complexity to the intensity of the others. In the *grands crus* the Bâtard-Montrachet is sweet-scented and creamy-textured with ample flesh adding opulence. Somehow, the oh-so-elegant Chevalier-Montrachet combines all these qualities, with added savour and citrus and a touch of almonds to push it beyond mere words.

Domaine Leflaive once was, should be, and hopefully will be again, the Côte de Beaune's white equivalent to the Côte de Nuit's Domaine de la Romanée-Conti in red. That day is closer now than when de la Morandière took charge. And that, though qualified, is an achievement worth heralding.

Try this: Puligny-Montrachet 1er cru Les Pucelles
I have always thought of this wine, when on song, as the quintessence of the Leflaive style – and the leading exemplar of *vrai* Puligny. The domaine is the largest owner here and though the wine may not carry the outright weight and conviction of, for instance, the Chevalier-Montrachet,

224 CÔTE D'OR

it possesses a grace and elegance that few can match. It is deceptively light, but this is a ballet dancer's lightness, there is extraordinary intensity, allied to refinement and finesse. From Leflaive, Pucelles is a *grand cru* in all but name.

Domaine Etienne Sauzet

11 rue Poiseul, 21190 Puligny-Montrachet
www.etiennesauzet.com; tel: +33 3 80 21 32 10

'I didn't know if we had enough freshness but we got a great surprise at the end of the élevage.' Benoît Riffault is speaking of the 2022 vintage and his words could be echoed by a legion of Côte d'Or producers, as the passage of time gradually embellishes the reputation of 2022, moving it into the top league of white vintages alongside such as 2014 and 2017 (the 2022 vintage in red is special too). Riffaut is one half of the management team at Sauzet, along with his wife Emilie Boudot, daughter of Gérard Boudot, grandson-in-law of Etienne Sauzet, who died in 1975.

We are conversing in the domaine's new tasting room, completed in 2023, which encompasses a magnificent view across the *village* Les Meix vineyard, towards the Montrachets on yonder hill. Sweep right from here and the panorama will take in almost all the Sauzet vineyards, which have been cultivated biodynamically since 2009, with certification in 2013. Nothing so clearly indicates the tiny scale of cultivation on the 'Golden Slope' as this view, which can be taken in with barely more than a glance.

Thus the journey the grapes make from vine to winery, after hand-picking into 40-kilogram panniers, can be measured in metres rather than kilometres. In the winery, they are sorted on a sorting table, 'we choose only what we can eat', and pressed with the stems, never crushed. A settling day in tank follows, before transfer to 228- and 350-litre barrels, up to 25 per cent new for the *grands crus*. After about 10 to 12 months they are racked back to tank for the second winter. The results of this deceptively simple approach are impressive.

Notwithstanding the quality of the *grands crus* the nine *premiers crus* that form the tasting room vista represent the bedrock of Domaine Sauzet. A quintet: lesser known La Garenne is fine and delicate with no lack of intensity; Perrières is precision-flavoured, unfolding into a savoury finish; Champ Canet is more structured, reserved with a hint of austerity and great tension; Folatières has greater weight on the palate, fuller in every respect; Combettes starts and finishes with a flourish, courtesy of a fantastic sweet–sour oscillation that would overwhelm were it not so well balanced. These are patrician Pulignys, there are no bumps on the palate, just a seamless flavour flow.

Try this: Montrachet grand cru

Sky-high expectations in advance of drinking a Montrachet are sometimes the precursor to a lukewarm reaction; a what's-all-the-fuss-about response can niggle on the margins. Not with the Sauzet Montrachet. This delivers the expected flavour symphony as it sways across an enchanted palate, caressing the taste buds all the way. It is seamless and polished, everything is present in balanced and boundless measure, there's not a hair out of place. Each sip reveals a variation on the original theme; melody becomes harmony as more and more flavours join the chorus. Though made from bought-in grapes on a long-term contract (the source is not revealed) this wine lacks for nothing. The cornucopia of flavours continues into a finish that lasts for fully thirty minutes.

CHASSAGNE-MONTRACHET

Chassagne-Montrachet may produce some of the most compelling white wines on earth, and delicious reds too, but as a village it possesses little of consequence, being strung out and largely centreless. It snakes along, rising and falling across the hilly terrain, with winding streets that narrow alarmingly in places, before petering out into vineyard on all sides. The wines of Chassagne and Puligny may differ subtly but the contrast between the villages themselves is striking. Chassagne boasts few of the grand houses that populate Puligny, nor its lavishly restored town square. Nor does it exude its hauteur.

The unprepossessing visual impact of the village should not be taken for a lack of vibrancy; a plausible claim could be made for Chassagne being the most exciting commune in the Côte d'Or right now. Coming across a poorly made Chassagne today brings with it a shock; it's not usual in a village replete with talented winemakers turning out wines of great purity and enduring appeal. Drinking them is a delight but keeping track of their makers is a different matter and could challenge a chess grandmaster. Many are related by marriage,

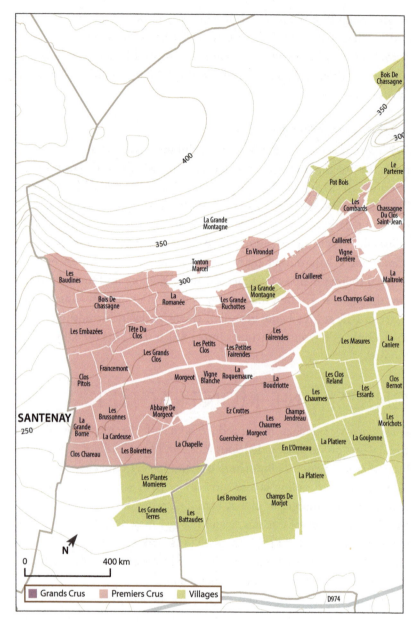

Map 12: Chassagne-Montrachet

their names interwoven and linked and complicated by the ubiquitous hyphen.

Chassagne's worldwide renown rests on the quality of its white wines, which are more linear and less rounded than Puligny's and if, ultimately,

THE VILLAGES AND PRODUCERS OF THE CÔTE DE BEAUNE

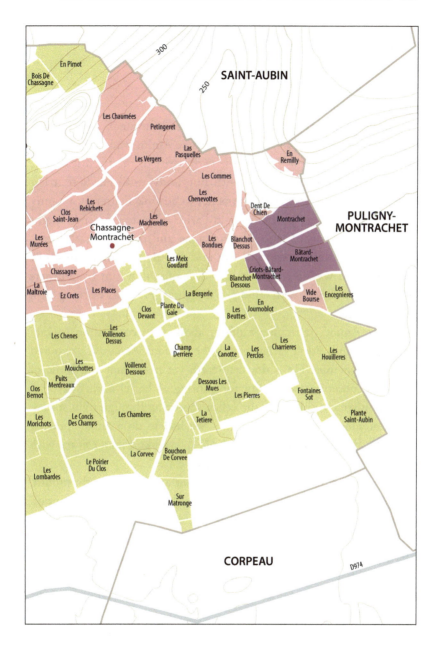

they do not bestow quite the same degree of satisfaction, and that is debatable, they probably deliver more by way of immediate thrill on the palate. A good *village* Chassagne-Montrachet is one of my favourite white burgundies.

Producers

Domaine Guy Amiot et Fils

13 rue du Grand Puits, 21190 Chassagne-Montrachet

www.domaine-amiot.com; tel: +33 3 45 63 81 31

'I have nothing crazy to tell you, we are conventional,' says Fabrice Amiot, great-grandson of Arsène and Flavie Amiot who ran a successful laundry business in Paris in the early years of the twentieth century. They had no connections with Burgundy, until Flavie sent her four children to Chassagne-Montrachet to be looked after by a nanny recommended by her postwoman in Paris. Thereafter Arsène bought property and vineyard land in Chassagne and one of the children, Pierre, started making wine in the mid-1930s. By this roundabout route the seed was sown for today's domaine, now run by brothers Thierry and Fabrice, with Thierry's daughter, Héloise, working in a part-time role.

Fabrice Amiot's refreshing candour about their conventional winemaking methods sits at variance with the passionate espousals favoured by some vignerons. He is down-to-earth yet never dull and some of his candid observations are not for quoting. A barrel tasting involves darting hither and thither in the cellar, squeezing through tight spaces, while catching Amiot's pithy asides: 'We had to do a lot of spraying in 2024 ...we don't claim to be organic ... we like a long élevage with bâtonnage ... chic we try but traditional we are.'

The results are white wines with amplitude and volume in the mouth, well removed from the lean and racy style that garners most attention today. Chassin, Damy and Tonnellerie Mercurey barrels are used, mainly 228-litre with some 300-litre, and tasting across the range of eight Chassagne-Montrachet *premiers crus* repeatedly confirms the house style, though it is not pursued at the expense of terroir differences: 'The winemaking for all the wines is the same; the differences come from the terroir,' says Amiot.

Les Macherelles could be a flag-bearer for the house style, plump and succulent, almost sumptuous. 'It has energy, power and length, with a yummy texture,' says Amiot, while its upslope neighbour, Les Vergers, delivers flavour pyrotechnics, dazzling the palate, 'it's got potential, it needs time'. Les Caillerets, just a few steps out of the winery, and of which Amiot owns over half a hectare, displays exotic tropical fruit, counterpoised by a streak of minerality. And then there is the

Some of Burgundy's greatest white wines emanate from the Demoiselles vineyard in Puligny-Montrachet

Montrachet itself, the two-parcel holding amounting to about one-tenth of a hectare and producing about two barrels. It is a multi-faceted wine, replete with opulent fruit and great depth of flavour. It would be a crime to drink it at less than a decade old.

Almost the entire production of 80,000 bottles, from 12 hectares, is exported to 40 countries, with a mere 5 per cent sold in France. Red wines make up about one-third of production, with a vigorous, dark-fruited La Maltroie 1er cru giving the lie to the commonly held belief that Chassagne-Montrachet *rouge* isn't a 'serious' wine.

Try this: Puligny-Montrachet 1er cru Les Demoiselles
Named after two ladies who owned the vineyard in the nineteenth century and who died without marrying, Demoiselles borders Montrachet and Chevalier-Montrachet, and in Amiot's hands is of *grand cru* quality. Choosing a Puligny from a domaine so deeply rooted in Chassagne might seem strange, yet the pure, seamless beauty of this wine cannot be ignored. So complete and rounded is the flavour that it seems churlish to parse it into fruit, acid and so forth. There isn't a hair out of place, so it is far better to enjoy the overall effect. 'Now I feel young again,' quips Fabrice Amiot as he tastes the 2023 from barrel.

Domaine Bruno Colin

6 place de l'Eglise, 21190 Chassagne-Montrachet
www.domaine-bruno-colin.com; tel: +33 3 80 24 75 61

There is more to winemaking than making wine. It may be trite to labour this self-evident truth but it is remarkable that many winemakers, wizards in the cellar, devote little thought and imagination to marketing their wines. Only a tiny number can rely on the wines selling themselves. Some plod through the necessary steps, while a few show a spark of imagination. One such is Bruno Colin.

At the 2024 edition of Les Grands Jours de Bourgogne (see p. 290) hundreds of vignerons vied for the attendees' attention – and Colin caught it by the simple expedient of pouring some older vintages from large format bottles. Not only were they eye-catching, they also illustrated how the wines develop over time. The steady stream of visitors to his stand attested to the wisdom of this easy move.

Unravelling the complicated interconnections of the Colin family calls for a sharp mind and a good memory, and would take a verbal treatise (it's easier by diagrammatic family tree). Suffice to say that Bruno is the brother of Philippe and that their vineyards came down

Bruno Colin at Les Grands Jours de Bourgogne 2024

the family line from their father Michel Colin-Deleger. Prime amongst them is the tiny parcel of Chevalier-Montrachet, sufficient to produce an average of one barrel per year of outstanding wine. The minute sliver is almost impossible to discern on the map, yet no such problems trouble the taste buds when graced with this nectar. First made by Colin in 2016, it is a wine of almost riotous intensity, while retaining remarkable grace, elegance and hidden power, with an opulent overlay adding complexity.

In truth the Chevalier stands well above the other Colin wines but that is testament to its quality and not any slackness on their part. Production divides 70:30 in favour of white, including an impressive array of Chassagne *premiers crus*. To pick two: Les Chaumées forms the north-west corner of the commune before the hillside turns into Saint-Aubin and the wine exhibits a lovely interplay between rich and lean notes; La Maltroie trumps it by way of greater intensity and length. Perhaps there is a touch of auto-suggestion when tasting Maltroie, however, for the picture window in the slick new tasting room, built in 2021, looks directly onto that vineyard.

The grapes are hand harvested and selected at the vineyard, crushed and pressed, with the juice going into stainless steel tank for two days' *debourbage*. Fermentation starts in tank before transfer to barrel, all 350 litres, from coopers such as Billon, François Frères, Cavin and Seguin Moreau. For his *grands crus* Colin uses two- to three-year-old oak, never new, and also utilizes a small number of the uber-trendy glass globes, liking the freshness they bring. The weather-ravaged 2021 Chevalier, of which he had 115 litres, saw no oak at all.

The red wines have seen considerable improvement over the last decade. Where once they could have a coarse edge today the fruit shines and the tannins no longer oppress. There is polish and refinement as well as vigour, best seen in the Chassagne-Montrachet Vieilles Vignes. Made from four parcels of old vines, averaging 50 years of age, this wine explodes with raspberry and cherry fruit, tingling and dancing on the palate, before departing in a lip-smacking finish. Currently, Domaine Bruno Colin is on top of its game and with Bruno's son, Gabriel, studying winemaking in Beaune the future looks secure.

Try this: Santenay 1er cru Les Gravières Rouge

Along with Clos Rousseau, Gravières is the largest *premier cru* in Santenay, at almost 24 hectares. At its north-eastern corner lies Clos de Tavannes,

232 CÔTE D'OR

which is a separate *lieu-dit*, though it labours under the *climat* name of Les Gravières-Clos de Tavannes. In Bruno Colin's hands the wine is less robust than some: the nose is crisp and fresh and leads into a palate defined by ripe red fruit laced by a dry bite of tannin. The wine is refined without being austere and one sip immediately suggests another.

Maison Róisín Curley

9 rue Chardonnay, 21190 Chassagne-Montrachet

info@roisincurley.com

Where to start? In no particular order, master of wine Róisín Curley is a qualified pharmacist, holds a master's degree in viticulture and oenology and is a micro-négoce winemaker in Burgundy. She is also diminutive and soft-spoken, as well as being self-effacing to a remarkable degree. She is always smiling, never confrontational, yet to have achieved all that she has speaks of well-hidden steel beneath that cheerful exterior. It would be all too easy to underestimate Curley.

Growing up in the west of Ireland Curley credits her mother with imbuing her with her love of wine: 'She enrolled in WSET courses herself and was a huge influence, encouraging and cementing my interest.' Pharmaceutical studies then took her to Aberdeen for five years, before a return to Ireland where she completed the WSET Diploma and then enrolled in the MSc course. That took her to Montpellier for one year, Geisenheim for another, followed by a year at Château Latour, working on her thesis, a comparative study of conventional, organic and biodynamic viticulture and winemaking.

All the while: 'I was trying to figure out a way that I could make my own wine.' An opening presented itself in Burgundy when, through some contacts there, 'doors opened for me'. In short, she rented some space, bought some grapes and made her first vintage in 2015.

Since then she has gradually established relationships with growers and also works through a *courtier* when sourcing suitable fruit, her core wines coming from Beaune in red and Saint-Romain in white. Though she does not work the vineyards she has some control over the fruit she receives, buying grapes picked into 12-kilogram boxes early in the day. For the whites a gentle pressing follows, with a 24-hour settling before the still-murky juice is transferred to barrel. They come from a selection of coopers: François Frères, Chassin, Taransaud, Remond, Cadus and Stockinger from Austria, in various sizes, including 350- and 500-litre vessels. She does no bâtonnage and assembles the wines back in stainless

THE VILLAGES AND PRODUCERS OF THE CÔTE DE BEAUNE 233

steel before bottling. The results match Curley's words: 'I am aiming for finesse and elegance'. These are not palate pounders.

For the red wines, after sorting, she may use anything between 30 and 80 per cent whole bunches: 'There is no recipe, not at all, I look and taste to decide.' Fermentation is in stainless steel, with pumping over favoured, though she may do some punch-downs. She tastes every day, 'I respect the science but I work very much on taste and feel.' Pressing follows, using a vertical press, then at least a year in barrel, probably more, before bottling.

Curley continues: 'I always wanted to make Beaune, I love the flavour profile of Beaune wines ... I don't want my mark to be on the wines.' She may not be leaving her mark but her style speaks of a gentle hand in the winery, suggesting a close fit between her practice and Beaune's potential. The wines are soft and rounded with abundant red fruits and discreet sweetness on the finish.

Curley has come a long way in ten years, now exporting to Ireland, Denmark, Canada, Sweden, Australia and other markets. She still divides her time between the family pharmacy at home and winemaking in Burgundy. And what of the future? 'I don't have the answer. If I owned vineyards I'd move here.'

Try this: Beaune 1er cru Blanches Fleurs

Blanches Fleurs (also Blanche Fleur) lies at the northern limit of Beaune, bordered by Clos du Roi on one side and the A6 Autoroute du Soleil on the other. Being only a tad over 3 hectares it is seldom seen and, in truth, does not sit high in the Beaune hierarchy. Yet Curley manages to imbue this wine with more than just the simple fruit one might expect. Added appeal – even in the capricious 2021 vintage – comes by way of gentle sweet spice. A light and lissom wine, delicate and transparent, not a heavy hitter. Very much in tune with Curley's stated style.

Domaine Fontaine-Gagnard

19 route de Santenay, 21190 Chassagne-Montrachet
www.domaine-fontaine-gagnard.com; tel: +33 3 80 21 35 50

As the Tour de France whipped past the Fontaine-Gagnard premises on Thursday 4 July 2024 a *stagiaire* from the domaine went amongst the spectators, offering samples of wine in paper cups, *sans* vintage and appellation. It was delicious and in a world where too much sanctimony attends the appreciation of wine it was a reminder that context is everything – allied to a kind gesture in this case.

234 CÔTE D'OR

Kind gestures notwithstanding, the wines of Fontaine-Gagnard are worth seeking out. When speaking of the increasingly fêted 2022 vintage Céline Fontaine, in charge at this domaine since 2007, notes that the flowering was good, there was a long veraison and an early harvest. Picking began on 26 August, which was six days later than in 2020, their earliest ever, and yields were good, though not excessive. It is clear she is happy with 2022, though looking at the bigger picture she expresses herself, 'especially worried' by climate change. The vines are being stressed and although 'Pinot Noir adapts better than Chardonnay', she sees a time, perhaps 20 years hence, when 'the highest terroirs may disappear', because of water shortage.

None of this should detract from the beauty of the wines, essays in harmony, replete with rich though never obtrusive flavours. Gentle insistence is their calling card. The domaine is a relative youngster and dates from 1985, when Richard Fontaine, previously an air force mechanic, married Laurence Gagnard and they established their domaine, initially with 4 hectares, since expanded to 12. Their daughter Céline is a safe custodian of the impressive array of vineyard holdings astutely assembled over the decades and now including a trio of *grands crus* – Le Montrachet, Bâtard-Montrachet and Criots-Bâtard-Montrachet.

Fontaine-Gagnard owns one-third of a hectare in tiny Criots (1.5 hectares), from which comes a wine of great penetration and length thanks to an intense, creamy citrus character. The Bâtard is not as full-bodied as some examples – it is elegant and poised, racy and long. The Montrachet, from a 0.08-hectare parcel, is exotic and colourful; waves of flavour come on and on in a seemingly endless stream of gustatory delight.

The wide range of *premiers crus* could be used to deliver a Chassagne-Montrachet masterclass. Les Vergers, on the slope looking across toward Montrachet, gets Fontaine's vote in the 2022 vintage: 'The soil is shallow and there is good acidity, this is my favourite 2022 with a lot of energy.' That said, the Cailleret and La Maltroie run it close, the former thanks to elegant insistence and the latter by way of a chalky, dry mineral bite and zippy acidity.

Though the white wines take centre stage at Fontaine-Gagnard it would be wrong to dismiss the reds as afterthoughts, especially the entry level Passetoutgrains. 'Burgundians have lost the taste for ordinary wines,' says Fontaine, a charge that could be levelled at many Burgundy lovers around the globe, now always seeking grandeur on the palate. But

THE VILLAGES AND PRODUCERS OF THE CÔTE DE BEAUNE 235

to ignore wines such as this crunchy-fruited delight is to miss out on top class quality without top class prices. Basic Burgundy is better than ever, no longer mean and sinewy. One sip of this Passetoutgrains should convince any doubters.

Try this: Chassagne-Montrachet 1er cru La Romanée
'Some people call it the sky slope,' says Céline Fontaine, referencing La Romanée's steepness and the dangers of working with tractors there. The vineyard is a *lieu-dit* of the *climat* La Grande Montagne, a large, catch-all area that includes vineyards and scrubland. Erosion of the slope means that the topsoil in the upper part is only 20 centimetres deep and the 70-year-old vines need replanting. Yet none of this takes anything from the quality of this wine; out of adversity comes beauty. Expansive and plump on the palate with distinct floral notes and subtler spice, the wine is remarkable for combining density and elegance, before resolving into a ringing finish.

Domaine Armand Heitz

1 rue du Château, 71150 Chaudenay
www.armandheitz.com; tel: +33 3 80 41 64 27

'I try to connect my work with nature,' says Armand Heitz as he lights a candelabra to add some gentle light to that coming from the crackling fire and the feeble winter light from the window in his tasting room, though 'cosy parlour' describes it better. He continues: 'We need to connect all the various agricultural endeavours again … 60 years ago there were cattle amongst the vines … vineyards need diversity.' For Heitz, making wine is not an isolated activity, ring-fenced as it were from other uses of the land to grow crops and raise animals to feed ourselves.

Heitz talks quietly but with notable insistence as he enunciates his philosophy and practices, which yield wines of a style that can be summed up in one of his pet phrases: 'We make soft extractions.' The red Chassagne-Montrachet 1er cru Morgeot is a case in point. Morgeot is a sprawling, catch-all *climat* that covers 58 hectares, and its red wines have a reputation for robustness over finesse but not this one. The grapes come from two *lieux-dits* – Les Petits Clos and Tête du Clos – and the wine is made with 100 per cent whole clusters, seeing no more than 25 per cent new oak, low toast. Heitz describes it as 'soft, elegant and refreshing', though beneath the restraint and balance lies some well-integrated power. As such, it might be considered a cousin of the Clos des Poutures (see below).

236 CÔTE D'OR

The white Morgeot marches to a similar beat, majoring in an incisive citrus cut with a saline snap on the finish. The Meursault La Barre delivers a fuller, more ample mouthful, though it is only with some age that the full panoply of flavours is released in these wines. The Meursault 1er cru Les Perrières 2015 drunk at eight years old had expanded from an initial linear profile to add plump fruit to the crisp, mineral framework. At entry level, the Bourgogne Blanc is something of an outlier, containing 10 per cent each of Pinots Blanc and Gris in a soft and easy package that could use a little more vibrancy on the palate; it tends to slide across without leaving much impression.

Armand Heitz is a man happy to chart his own course, guided by firmly held, and repeatedly expressed, beliefs. 'There is no point making wines I don't like,' he concludes.

Try this: Pommard 1er cru Clos des Poutures monopole

The *clos* is a 0.66-hectare enclave within the 4 hectares of Les Poutures, which abuts the village on its southern side. Might this be considered a Pommard in name only? Yes, if your taste is for old-school Pommard and you want some heft and ballast in your wine. No, if you lean towards a more modern style that walks on lighter feet. Armand Heitz explains: 'I wanted to make a wine I appreciated. It didn't match people's impression of Pommard, so everybody says to me, "It is not a Pommard."' To Heitz's thinking it is, 'more authentic'. Regardless of authenticity the wine cannot be faulted for purity of flavour and, judged by what is in the glass and not on the label, it delivers a soft-textured, mouth-watering wine with juicy, sweet fruit and a gentle whisper of acidity and tannin.

Domaine Alex Moreau

21 route de Santenay, 21190 Chassagne-Montrachet
Tel: +33 3 80 21 33 70

'Family', said with an inscrutable smile and a wistful shrug, is all Alex Moreau offers by way of reply when asked about the reasons behind the split that saw Domaine Bernard Moreau divided between him and his brother, Benoît, after the 2020 harvest. Bernard is their father and the brothers joined him in 1999 (Alex) and 2002 (Benoît). *Père* retired in 2012 and by the time of the split the domaine's reputation for white wines of grace and harmony that never failed to satisfy was amongst the best. At the time the domaine encompassed 10 hectares of their

THE VILLAGES AND PRODUCERS OF THE CÔTE DE BEAUNE 237

own vineyards along with another 5 hectares farmed *en fermage*. Alex Moreau now has charge of 5.7 hectares, which is, he says, 'better than nothing'.

He hasn't lost his touch. The Bourgogne Blanc is an exceptional wine – a glossy overlay on a firm citric core delivers a beautiful lemon curd character on the palate: 'I pay attention with Bourgogne Blanc, it needs it and it makes sure everybody can enjoy wine, it is very important to have good quality, it touches everybody.' For the Chassagne-Montrachet *village* two-thirds of the fruit comes from the Puligny side of the commune, the remainder from plots close to Champs Gain and Morgeot. The former brings freshness, the latter structure. It is rich and fresh by equal parts, with a soft texture wrapping a firmer, mineral core: 'Wine is made to be drunk, not just tasted, I like salinity,' says Moreau.

He also talks about the challenges of climate change: 'Harvest is more stressful now, choosing the date to pick. I have seen 25 harvests but now the window to pick is narrower than ever. If you want to preserve your style you have to get it right. I am happy with 2022 now but didn't expect this freshness during harvest.' And he now prunes later, 'even by a few days', doing the *premiers crus* last and not first as previously, in an effort to lessen the risk of spring frost destroying the young buds.

For the red wines Alex Moreau has changed his style radically from what he did 15 to 20 years ago: 'I now do the opposite to what I once did, I went for much more extraction then.' That change is most evident in the Chassagne *monopole* La Cardeuse, the 2007 vintage of which snapped on the palate in youth, only settling after a dozen years in the cellar, whereas the 2017 was charming from its fifth birthday on. Made with 35 per cent whole bunches, the wine is less stridently flavoured, though still delivers a palate of concentrated dark fruit and firm tannin. Unfortunately, after the split this is no longer a *monopole*, though it will be interesting to compare the brothers' interpretations of it.

Try this: Chassagne-Montrachet 1er cru Les Chenevottes
The 11-hectare Chenevottes faces directly east and sits at the bottom of the gentle slope that rises towards Saint-Aubin. The name probably derives from the fact that hemp was once cultivated here. Alex Moreau picks this vineyard on the first day of harvest, 'as maturity comes fast and if I miss the window I will get 14.5 per cent alcohol and that is not my style.' There is bountiful fruit, juicy acidity and rich depth, yet the overall impression is one of elegance, finesse and restraint. A mild

238 CÔTE D'OR

mineral streak means the wine is never heavy or overbearing, satisfying from first to last sip.

Domaine Marc Morey

3 rue Charles Paquelin, 21190 Chassagne-Montrachet
www.domaine-marc-morey.fr; tel: +33 3 80 21 30 11

An easy way to gain a quick measure of a winemaker's ability is to taste their most basic wine, working on the premise that to make a good wine at this level takes real talent, whereas if they cannot turn out a decent *grand cru* they are in the wrong business. It is a litmus test that this domaine sails through, thanks to the sensitive winemaking touch of Sabine Mollard, granddaughter of Marc Morey, who started here with her father, Bernard, in 2003. He retired in 2006, though the first vintage she made entirely on her own was 2010 (look for the corks branded 'by Sabine'). Carrying out this exercise with her Aligoté brings enduring gustatory delight, by way of delicious fruit and singing purity. The grape's signature perk is there too, but in civil measure, it is not the spike of old that sent drinkers scurrying for the crème de cassis.

Mollard is not one for voluble expositions of her philosophy or practice. A sample is poured, discussed without undue elaboration, and you move on to the next one. The wines do the talking and they speak in well-composed, restrained, elegant tones, with a succulent overlay in some cases. The grapes are hand harvested, selected in the vineyard and given a quick pressing, after which the juice spends two days in tank at 12°C for the *debourbage*. Thereafter the temperature is allowed to rise and fermentation starts in tank, at the mid-point of which the must is transferred to barrel – from coopers Chassin, François Frères and Tonnellerie du Val de Loire – for about 10 months. Thereafter it is back into tank for a second winter with bottling early in the new year. The results at *premier cru* level are impressive: a beautifully balanced Cailleret, elegant and long; a crisp Les Vergers; and a succulent Chenevottes, rich and ample by comparison.

For the red wines – principally a Chassagne *village* – the grapes are partly de-stemmed, with about 30 per cent whole bunches retained, and these are then layered into the tank, with plenty of *remontage* and very little *pigeage* during fermentation: 'I like fruity wines without too much extraction,' explains Mollard. This wine is abundantly fruity and pure, giving emphasis to Chassagne's gradually rising reputation as a *rouge* source.

THE VILLAGES AND PRODUCERS OF THE CÔTE DE BEAUNE   239

If there is a cloud on the horizon at Domaine Marc Morey it is the lack of any clear line of succession after Sabine Mollard. When asked about the future, she simply replies: 'I don't know. I don't think about this now.'

Try this: Chassagne-Montrachet 1er cru En Virondot

Q. When is a *monopole* not a *monopole*? A. When it is Morey's En Virondot. The domaine is the only producer of this wine and owns all but a garden-sized plot of the vineyard, whose grapes it buys, which are included in its *village* Chassagne. The remaining 2 hectares comprise the domaine's largest holding, the vineyard sitting right up at the treeline, at over 300 metres altitude. The wine is beautifully composed; strength and delicacy compete for the palate's attention, the first coming from rich fruit, the second endowed by a streak of minerality.

Domaine Vincent & Sophie Morey

3 Hameau du Morgeot, 21190 Chassagne-Montrachet
www.morey-vs-vins.fr; tel: +33 6 50 97 94 55

The scourge of Covid threw up some surprising benefits, a modest one of which is the new tasting room that Vincent Morey built during lockdown – in a slick and modern idiom, with stainless steel shelving, bright orange seating and not a cobweb in sight. Vincent succeeded his father Bernard in 2007 and today the 20-hectare domaine includes vines from the family of his wife, Sophie, whom he married in 2002. Their vineyards lie mainly in Chassagne with other holdings in Santenay, Maranges and Saint-Aubin.

Vincent Morey's philosophy does not chime with the current fashion for early harvesting to retain acidity, which means the house style majors on fuller, less incisive flavours: 'I don't harvest too early, maturity is most important for me, I belong to an older generation. The younger generation tends to harvest too early, looking for acidity instead of real maturity. In 2022 I started picking on 1 September – most of my neighbours had finished by then.' Thus the acidity which is front and centre in many white burgundies today lies deep within his wines, a hidden framework, not an exoskeleton.

The Bâtard-Montrachet, from a holding that forms a narrow strip running almost the width of the vineyard on the Chassagne side, amounting to a mere tenth of a hectare, is the perfect exemplar of the house style, polished and plush, deep and satisfying. It leads a cohort

240 CÔTE D'OR

of similarly styled *premiers crus* including Les Baudines, Morgeot and Cailleret from Chassagne, with an outlier from Puligny in the shape of a creamy and crisp La Truffière.

Production divides 50:50 between white and red, with much of the latter coming from Chassagne *village* and several Santenay *premiers crus* such as Passetemps, Les Gravières and Beaurepaire. The Chassagne is assembled from four plots and is a sumptuous and broad-flavoured wine, while the Beaurepaire majors on sweet, liquorice fruit with a note of spice on the finish. It is a rich, lavish wine, easy quaffable in youth but well able to handle five to seven years in the cellar.

Perhaps the most appealing thing about the Vincent & Sophie Morey wines is that they march to a beat that is a little out of step with the current zeitgeist, bringing variety in a world threatened by homogeneity. For that they should be celebrated.

Try this: Chassagne-Montrachet 1er cru Les Embrazées
Though Vincent Morey labels this wine 'Embrazées' the more usual spelling is 'Embazées' – and, confusingly, other versions can also be found. This *lieu-dit* sits at the southern end of Chassagne, on the boundary with Santenay, and there is a hand-in-glove fit between vineyard potential and the ample house style. Vincent Morey is by far the largest owner here, with almost 4 hectares. In addition to the expected opulence, there is a more complex and less easy to pin down herbaceous quality to this wine. It is a will-o'-the-wisp flavour inflection that adds a *sauvage* character, thus endowing the wine with enduring attraction.

Domaine Jean-Claude Ramonet

4 place des Noyers, 21190 Chassagne-Montrachet
www.chassagne-montrachet.com/en/producteurs/domaine-ramonet-2/; tel: +33 3 80 21 30 88

There has been much change at Domaine Ramonet in the last decade. Prior to 2014 brothers Noël and Jean-Claude, grandsons of founder Pierre, worked side-by-side, running the domaine jointly. Under their stewardship the reputation soared worldwide, driven by wines of notable character and longevity, most especially the whites. After reorganization in 2014 Noël stepped back from day-to-day involvement, a change subtly indicated by the wines being labelled as 'Jean-Claude Ramonet' since then. More recently, Noël's sons, Michaël and Pierre-François,

THE VILLAGES AND PRODUCERS OF THE CÔTE DE BEAUNE 241

have started to make some wines in their own name, reclaiming their share of the vineyards. Jean-Claude has also been joined by his daughters, Anne-France and Clarisse, so the fourth generation is now firmly embedded at Ramonet.

To make good the loss of vineyards to his nephews Jean-Claude has, for some years, been buying or renting other parcels of vines, the first of which was Pernand-Vergelesses Les Belles Filles (not named for his daughters) from which he produces both a red and a white. Since 2022, a sliver of Corton-Charlemagne has been added, a sumptuous and savoury, creamy-textured wine that has few rivals on the Hill of Corton. At a less exalted level, and beyond the Côte d'Or, a magnificent Mâcon-Péronne is a top-quality entry-level wine.

Where it was always Jean-Claude who hosted tastings you are now as likely to be met by Anne-France, accompanied by her dog Nabu. Speaking of recent vintages, most particularly the delights of the underrated 2021 red wines, she says: '2022 is always charming; with 2021 you have to be a connoisseur to appreciate this kind of wine, it is true Pinot Noir. It is not too heavy, not too strong. It is fresh with lots of salinity. I really like the 2021 reds.' Which draws attention to the fact that, overshadowed by the renown of the white wines, the Ramonet reds remain slightly below the radar, delighted in by those who know them but only gradually being discovered by others.

The whites do deserve their centre-stage position, however, and, as with the reds, the 2021 vintage was a triumph at Ramonet, the lack of amplitude in the wines leaving them more 'visible' on the palate. After pressing, the juice goes to tank where fermentation starts without any settling of the solids. It is then transferred to barrel to finish fermentation and given about one year's élevage, followed by three months back in tank before bottling. The barrels come from five coopers: Damy, Rousseau, Seguin Moreau, Chassin and Tonnellerie Mercurey, and percentages of new oak range from 10 to 15 for *village* wines, through 20 to 30 for *premiers crus*, up to 50 for the *grands crus*.

A Chassagne-Montrachet masterclass comes by way of tasting through the eight *premiers crus*, all of which bear the Ramonet 'stamp' as it were, a signature *sauvage* quality that is both intriguing and elusive, especially when trying to capture it in a tasting note. Picking a favourite is a futile task, though three standouts are La Boudriotte, Clos des Caillerets and Les Vergers. The first fine and elegant, the second deceptively light at first but with great persistence, the third exotic and expansive with

242 CÔTE D'OR

succulent, ripe fruit. More than most, the Ramonet wines, in both colours, reward extended cellaring – at 20 years of age the Morgeot 1er cru 2004 Blanc was still lime-yellow to the eye and equally fresh and vibrant on the palate.

Try this: Montrachet grand cru

A tasting at Ramonet invariably ends with Montrachet, in which vineyard the domaine owns a 0.25-hectare slice that runs east–west on the Puligny side of the commune boundary, though this will now be split between the fourth generation. At its best this is a wine of extraordinary breadth and depth, delivering a rolling wave of flavour whose components jostle for attention on the palate. It's an exotic concoction of abundant fruit and savoury notes, all leavened with the domaine's signature herbaceous whiff. For minutes after tasting the palate hums with residual flavour that eventually fades away slowly.

SAINT-AUBIN

As with Auxey-Duresses and Saint-Romain, a visit to Saint-Aubin necessitates a diversion away from the côte's heartland. Travelling from Chassagne, a long left-hand sweep on the D906 takes you past a sliver of dwellings that constitutes part of the hamlet of Gamay, with Saint-Aubin a little further on. Its out-of-sight location may once have kept it out-of-mind but not any more. Today, it can be safely said that Saint-Aubin's best white wines stand compare with the big three of white burgundy: Meursault, Puligny-Montrachet and Chassagne-Montrachet. No longer do they attract the pejorative 'useful'.

If there is a fly in the ointment it is the preponderance of *premiers crus* – a dubious distinction it shares with Beaune (*village* vineyards amount to 80 hectares, *premiers crus* are almost double that). Such a top-heavy arrangement might raise Saint-Aubin's profile but it does nothing for the long-term reputation. It does not bestow status and drawing attention to it is a vacuous boast. What does raise the reputation are producers such as the triumvirate profiled below. In their hands a *premier cru* is a wine of distinction and character, well worth seeking out and well worth the extra money. In terms of consumer recognition it does the wines of Saint-Aubin no harm that these vignerons have names that are also associated with Chassagne, and indeed they all have vineyard holdings there. In Burgundy, where the golden buying rule is to get the name of the producer first, this constitutes a welcome fillip.

Producers

Domaine Jean-Claude Bachelet et Fils

15 rue de la Chatenière, Hameau de Gamay, 21190 Saint-Aubin
www.domainebachelet.fr; tel: +33 3 80 21 31 01

It can be said with certainty that the Bachelet family traces its roots in the region of Saint-Aubin back to 1648. The accuracy stems from considerable genealogical research done a few years ago to coincide with the visit of Chilean president Michelle Bachelet, an ancestor of whose emigrated to South America in the nineteenth century. (There is a Parc Michelle Bachelet in Chassagne-Montrachet, opened by the president on a visit in 2009.)

Notwithstanding this ancient lineage the domaine as constituted today takes its name from the eponymous Jean-Claude Bachelet, who died in 2020. He had been in charge since the age of 20 after his father's sudden death, and today his sons, Benoît and Jean-Baptiste, have succeeded him. They were not subjected to the same sudden imposition of responsibility as *père*, having worked alongside their father for many years, Benoît since 2000 and Jean-Baptiste since 2005.

Domaine Bachelet started converting to biodynamics in 2012 and completed the transition in 2016, though the first certified vintage was 2023: a reluctance to deal with the necessary paperwork prompting the delay. Production from the 10 hectares, spread across Saint-Aubin, Chassagne-Montrachet and Puligny-Montrachet, divides about 75:25 in favour of white wines, whose signature style derives from an approach to vinification that deviates slightly from the early harvest and less oak norm today.

Regarding the picking date, Benoît emphasizes that he harvests neither early nor late, instead seeking complete maturity, even if that means more sugar, adding that in sunny years some bitterness in the grapes will compensate for the low acid. In the winery an extended élevage that sees the wines spending two winters in barrel, followed by some time in tank before bottling, suggests that the style might major on opulence over finesse.

That is not the case, the Bachelet wines are full flavoured but they carry their weight easily, as confirmed by a trio of *premiers crus*, one from each commune: Saint-Aubin La Chatèniere, Chassagne-Montrachet

244 CÔTE D'OR

Les Macherelles and Puligny-Montrachet Sous le Puits. The Chatenière vineyard sits directly opposite the winery and the wine, broad and lavish at entry, closes like an iris to a precise finish. The Macherelles delivers a great chorus of fruit that resolves into an echoing finish, while the Puits is the fullest of the three, the fruit burnished by some notes of butterscotch and shot through with mild spice. Riper picking and longer élevage are doing nothing to harm the Bachelet wines.

Try this: Saint-Aubin 1er cru Le Charmois

The name 'Charmois' derives from 'Charmes', meaning uncultivated land, a possible reference to a time when the land lay fallow due to the difficulty of working the relatively steep slope. Bachelet owns a little less than 1 hectare in this east-facing vineyard that lies on the boundary with Chassagne-Montrachet and looks across towards the Montrachet *grands crus*. Vine age is a good 30–35 years and they produce a wine of notable purity, derived from crisp acid and juicy fruit. Quite pointed on entry, the wine unfolds as it crosses the palate and departs with a warm glow on the finish.

Domaine Marc Colin

9 rue de la Chatenière, 21190 Saint-Aubin
www.marc-colin.com; tel: +33 3 80 21 30 43

Twenty-four carat consistency, with none of the safe and sound downside implied by that designation, is Domaine Marc Colin's calling card. And this is consistency at a high, sometimes exceptional, level. Right across the range the wines deliver compelling flavours, from humble Aligoté to patrician Montrachet. In some respects it is that basic Aligoté that signals the quality to be expected from Marc Colin – it takes real skill to craft excellence at this level – if you cannot do it with Montrachet you are in the wrong game.

Marc Colin is now retired and his four children, Pierre-Yves, Damien, Caroline and Joseph, have succeeded him, though Pierre-Yves struck out on his own 20 years ago and has since established a stellar reputation from his base in Chassagne-Montrachet, while Joseph left the fold in 2017 and now operates from a cellar only a stone's throw away, leaving Damien and Caroline in charge here. Production is almost exclusively white, with just 5 per cent of the holding planted to Pinot Noir. Montrachet and Bâtard-Montrachet are produced in minute quantities and sit at the top of the vineyard holdings, which also include a splendid range of Chassagne-Montrachet and Saint-Aubin *premiers crus*.

After harvest whole bunches are pressed and the juice goes directly to barrel with no settling, for about 10 months with no bâtonnage because, as Damien explains, 'we want freshness and precision'. Modest percentages of new barrels are used, sourced from a quartet of coopers: François Frères, Chassin, Cadus and Doreau from Cognac. The Bourgogne sees 5 per cent, rising to 10 for *village*, and ranging from 15 to 30 per cent for the Saint-Aubin and Chassagne *premiers crus*. No new barrels are used for the *grands crus*, given that they are produced in such minuscule measure. Élevage is completed in stainless steel for six to eight months prior to bottling in February or March and sealing with DIAM closures. The extra cellar space created by Joseph's departure allows for longer élevage – a somewhat paradoxical development, as the departure of a sibling, with their share of the family vineyards, is usually seen as a negative.

By these means Damien Colin aims to retain 'freshness and finesse', while welcoming the high proportion of old vines in his vineyards: 'Vine age is important for density and definition in the wines.' One such is the Puligny-Montrachet Les Enseignères whose 75-year-old vines yield about three barrels of citrus-infused wine, precise and perky with an enduring tingle on the finish. The Saint-Aubin 1er cru En Remilly makes a bigger statement, broader on the palate though not as fine. Meanwhile, its neighbouring *premier cru*, La Chatenière, displays a classier profile, polished and reserved, clean and pure.

Attractive though that trio is they are discreetly outclassed by the Chassagne-Montrachet 1er cru Les Caillerets, a wine that reflects the stony slope of its origins in a fresh and firm flavour palette, intense and incisive thanks to a racy streak of acidity and an almost tannic mineral bite. All it needs is patience and time.

Try this: Saint-Aubin 1er cru En Montceau

En Montceau is a *lieu-dit* of the *climat* Les Champlots and faces almost directly west at over 300 metres elevation. The Colin holding is a generous 1 hectare, from which is produced a wine of compelling class and composure, in many respects the domaine's flagship, notwithstanding the attractions of the *grands crus*. This wine hints at their grandeur without their bank-breaking prices. In youth it is firm and forceful, suffused with mineral intensity. Only after five to seven years, in a great vintage

246 CÔTE D'OR

such as 2017, does the austerity yield to more charming nuances of succulent fruit, with a whiff of truffle adding complexity. It is polished and pure, perfumed and elegant. A winner.

Domaine Hubert Lamy

6 rue du Paradis, 21190 Saint-Aubin

www.domainehubertlamy.com; tel: +33 3 80 21 32 55

How does a domaine achieve cult status? There must be an undeniable level of quality but that is not enough. Allied to it, and less easily definable, the personality and ability of the winemaker must set it apart, giving it a mark of distinction that others lack. In the case of Domaine Hubert Lamy that person is Olivier Lamy, behind whose smiling countenance lies a level of unalloyed intensity that few possess. His dedication to viticulture is unmatched – though I did a double take once when I spotted him tending his vines in Criots-Bâtard-Montrachet with a small chainsaw – cutting out rotten wood from some vine trunks as it transpired.

Though the Lamy family have tended vineyards for centuries the domaine of today is relatively young, founded by the eponymous Hubert in 1973. His son, Olivier, joined him in 1995 and in the decades since has expanded the vineyard holdings to 18.5 hectares, producing over 100,000 bottles annually. White wines account for 80 per cent of production and it is on these that the reputation for supremely fine, yet notably intense, wines has been built. For wines of purity, precision and poise Lamy is the apogee.

The secret is high-density planting, of which Olivier Lamy might be considered the high priest. Fourteen thousand vines per hectare is the baseline, all the way up to a scarcely credible 30,000 in the Derrière chez Edouard *climat*. Tending to vineyards thus planted is akin to self-imposed slavery, yet density on the ground delivers concentration on the palate, and then some. Concentration can be a blunt instrument unless deftly crafted, which in the case of Lamy wines it certainly is.

The wines drink beautifully when young – which is a paradoxical drawback – for temptation must be resisted if they are to be seen at their best when maturity adds a gloss to the flavour, imbuing them with a textural smoothness that speaks of true class. At 10 years old the Saint-Aubin 1er cru En Remilly 2014 was still bright to the eye, and bright on the palate too, the linear intensity of youth softened and broadened by a decade in the cellar.

That cellar has been extended since the first edition of this book and now features an end wall of bare rock, which provides the setting for an impromptu geology lesson as Olivier shows visitors around. Since 2017 also, the style has perhaps softened a little, the piercing minerality is still there, now joined by a whiff of tropical fruits to yield slightly less chiselled wines than previously. More than almost any others these are white wines that cry out for decanting – and don't be shy about letting them cascade and foam when doing so. With young vintages it is either that or a long spell in the cellar, to coax enduring delight from them.

Try this: Saint-Aubin 1er cru Clos de la Chatenière
Whether the name derives from *châtaigne*, chestnut tree, or not, is open to debate, though the quality of this wine is not. The vineyard runs to 8.5 hectares in a narrow strip that roughly follows the 300 metre contour line, looking south and west from a relatively steep slope close to the hamlet of Gamay. The wine runs slightly counter to the Lamy stereotype; the mineral core of pure acidity is there but it is fleshed out by abundant fruit, given a polish as it were. It is seamless and harmonious, with wonderful intensity and length.

SANTENAY

While it may not possess the stellar vinous reputation of its near neighbour, Chassagne-Montrachet, Santenay counts as one of the most hospitable of all the Côte d'Or villages and it is also considerably bigger than its neighbour. Travelling the vineyard route from Chassagne, Santenay is signalled by the windmill that sits above the road in the Beauregard vineyard, its sails stationary no matter what the breeze. The wonder is that there are no *moulin*-inspired *lieux-dits* surrounding it: 'Dessus du Moulin' and 'Dessous du Moulin' seem apt. Santenay is a two-part village, comprising Le Bas, close to the railway, and Le Haut, a smaller uphill annex that sits below the Montagne des Trois Croix, whose 500-metre summit is surmounted by three crosses, giving it a Calvary-esque appearance.

Until recently Santenay was the Côte de Beaune's country cousin, rustic tweed to refined satin. Its reputation was for solid reds that could add ballast to a merchant or restaurant list at the lower end of the price scale. Times have changed, and thanks to the efforts of a talented clutch of producers the wines have lost their rustic edge, shedding coarseness

248 CÔTE D'OR

for finesse, four-square flavours for juicy fruit. In this respect it is mimicking the recent rise to acclaim of Saint-Aubin. Pinot Noir planting still exceeds Chardonnay but the whites now offer, in some cases, superb quality at prices that are more attractive than those commanded by wines of equivalent quality from Chassagne or Puligny. It is in Santenay that the Côte d'Or begins to swing away from its roughly north–south orientation to the point where some of the vineyards are facing due south on rugged slopes surrounding the thirteenth-century chapel of Saint-Jean-de-Narosse.

Producers

Domaine Bachey-Legros
12 rue de la Charrière, 21590 Santenay
Tel: +33 3 80 20 64 14

Brothers Lenaïc and Samuel Legros currently have charge of this 20-hectare domaine, having worked alongside their mother, Christiane Bachey-Legros, for many years. Their holdings are spread across three communes: 12 hectares in Santenay, along with 3.5 each in Maranges and Chassagne-Montrachet, as well as 1 hectare of *régionale*. Production divides 70:30 in favour of red wines and it is principally upon those that the domaine's reputation rests. A small range of négociant wines is also produced, from Puligny-Montrachet, Meursault and Saint-Aubin.

For the red wines all the fruit is de-stemmed and only *remontage* is utilized during the vinification, which lasts between one and one and a half months. Samuel Legros explains: 'With old vines we don't need to do extraction, they give us the concentration. *Pigeage* was always too hard for the tannins, so we stopped.' Fermentation is in stainless steel, followed by élevage in standard barrels, 20–30 per cent new. The reds are dark and substantial with some Côte de Nuits gravitas and notable complexity in the best examples, thanks to impressive vine age across many of the vineyards. In fact, every one of the nine red wines is labelled *Vieilles Vignes* – a fully justified designation in Bachey-Legros' case.

The Chassagne-Montrachet Les Plantes Momières is a relative youngster, with vines dating from 1946. The wine is vigorous and concentrated when young, yet well rounded and, in a vintage such as 2015, it ages into sweet leathery splendour with a well-integrated tannic rasp. (It is also something of a hybrid vineyard, given that it lies within the commune of Remigny but may be labelled as Chassagne-Montrachet.) The

Santenay Clos des Hâtes claims vines from 1935 and Bachey-Legros owns a substantial 1.8 hectares in this vineyard that laps right up to the houses of the village. The wine is firm and strong, yet the bountiful fruit ensures immediate youthful appeal. Daddy of them all is the Santenay 1er cru Clos Rousseau Les Fourneaux, some of whose vines date from 1914, with the remainder from 1925. The wine is inky-rich and mouthfilling, substantial and deep; all it lacks is a steaming serving of *boeuf Bourguignon*.

The 2022 vintage, which was a success across the Côte d'Or, was particularly so at Bachey-Legros, delivering structured yet not too heavy wines: 'It is a little bit more Burgundian compared to 2018 or 2020, there is freshness and acidity,' says Samuel Legros. And, given that they are not afraid of innovation at Bachey-Legros, would Samuel consider using DIAM closures? 'No, no, no. We don't like them.'

Try this: Chassagne-Montrachet 1er cru Morgeot Les Petits Clos Vieilles Vignes
The name is a mouthful and so is the wine, though in the best sense. Notwithstanding Bachey-Legros' reputation for red wines this is the domaine's prize holding: 2 hectares in one plot towards the top of the vineyard, whose vines were planted in 1950. They yield a wine of real character, lively on the palate with no hint of lushness, plenty of fruit but also a nice saline undertow to give enduring attraction. It neatly avoids the heavy-hoofed trap that some Morgeot wines fall into.

Domaine Evenstad

1 place du Jet d'Eau, 21590 Santenay
www.domaineevenstad.com; tel: +33 3 79 35 00 35

The story starts in 1989 when Americans Ken and Grace Evenstad founded Domaine Serene in Oregon, focusing initially on Pinot Noir, followed by Chardonnay in 1997. Fast forward to 2015, when they acquired Château de la Crée in Santenay, and forward again to 2021 when they established Domaine Evenstad in the heart of the village, having bought the former Prosper Maufoux headquarters, while simultaneously taking control of Domaine Christian Confuron in Vougeot. Unfortunately Ken Evenstad died shortly afterwards but Grace continued with the project of establishing a well-founded and fully-functioning domaine in the Côte d'Or.

250 CÔTE D'OR

Much work was needed to bring the Santenay premises up to scratch and for a couple of years the village played host to a legion of construction workers as a massive renovation programme progressed at a sedate pace, finally drawing to a close in 2024. The once tired façade, which occupies almost one side of the square, is now trim and bright and, inside, the cellars are similarly impressive – and visitor friendly.

Thankfully the wines match the ambition of the premises. The maiden vintage was the well-regarded 2022, so luck was on Evenstad's side, though luck plays no hand in the vinification. As explained by Marc Magnat, manager in charge of day-to-day operations, the grapes are hand harvested, with some sorting in the vineyard and then on a vibrating table at the winery. Fermentation, utilizing some whole bunches depending on vintage, is in stainless steel with a combination of pumping over and punching down. Élevage of 12–18 months is principally in 228-litre *pièces*, about 30 per cent new, with a small number of currently fashionable glass globes to bring some freshness to the final assemblage.

The fruit comes from the former Confuron vineyards, comprising 6 hectares in the Côte de Nuits, including a slice of Clos de Vougeot and a garden-sized plot in Bonnes Mares, good for about one barrel of wine per year, though you don't have to go to the top of the tree to find a satisfying wine from the Evenstad range. The Côte de Nuits-Villages Les Retraits, which borders the Nuits *premier cru* Clos de la Maréchale, delivers crunchy fruit and crisp tannin, while the Nuits-Saint-Georges Les Longecourts, which neighbours Les Saints-Georges, majors on sweet, plump fruit and mild spice.

It is in Chambolle-Musigny, however, that the Evenstad wines really sing, most particularly in Les Feusselottes (see below) but also the *village* Les Condemennes, a downslope, 5-hectare vineyard that lies close to Vougeot. It is lacy and delicate but not frail, marked by splendid intensity and fresh length. As Magnat puts it: 'If you combine intensity with acidity you have a powerful wine, still elegant ... but if we lose one or the other we are dead.' Based on the 2022 vintage Evenstad is not losing either.

Try this: Chambolle-Musigny 1er cru Les Feusselottes
Les Feusselottes lies below the village, almost equidistant between Chambolle's two *grands crus*, Musigny and Bonnes Mares, and is easily found thanks to its surrounding the graveyard on three sides. The

Evenstad holding is comparatively large, almost a hectare, producing about 4,000 bottles on average. A beguiling perfume sets the scene for a palate of intense red fruits shot through with mild spice, all supported by a crisp tannic undertow. The 2022 will serve well as a flag carrier for Evenstad's maiden vintage, giving promise of even better things to come.

Domaine du Château Philippe le Hardi

1 rue du château, 21590 Santenay

www.philippelehardi.fr; tel: +33 3 80 20 61 87

What was once the 'Château de Santenay' has now been rebranded in long-winded fashion as 'Domaine du Château Philippe le Hardi'. It is but the latest chapter in a history that dates back to the fourteenth century, when the building itself was probably owned by Philippe le Hardi, Duke Philip the Bold, he of the 1395 decree that banned the planting of the 'very bad and very disloyal' Gamay grape. The château is one of the most eye-catching in the Côte d'Or, easily spotted thanks to its expansive polychrome roof, though at the end of 2024 one gable end was disfigured by temporary buttressing to guard against subsidence, while a more lasting solution was sought.

That glitch apart, the recent news from these impressive, though previously slumbering, premises above the village of Santenay has all been positive. A new broom in the person of *directeur général*, Jean-Philippe Archambaud, has been in place since 2019 and in that time he has effected a complete turnaround in the winemaking philosophy and practice. For years the wines trundled along below the radar, serviceable, but well out of kilter with the grandeur of the premises. That disconnect has largely been rectified and the change in winemaking can be summed up in a sound bite: *pigeage* is out, *remontage* is in. Rather than whacking flavour from the grapes as heretofore, the approach now is to coax it, in search of intensity rather than outright concentration.

The year 2019 also saw the purchase of the 8-hectare Domaine François Perrot in Gevrey-Chambertin. This gave an important toehold in the Côte de Nuits, including two long, thin slices of Clos de Vougeot, in the south-eastern quarter of the vineyard, amounting to 0.85 hectares, from which the newly minted domaine's flagship wine is now produced. The purchase brought the total holdings up to nearly 100 hectares, though about 60 of those lie outside the Côte d'Or, centred on Mercurey.

Conversion to organic farming was initiated in 2021, with certification achieved from the 2024 vintage. That year also saw a brand new

252 CÔTE D'OR

set of wooden vats from Tonnellerie Rousseau used for the first time, though their intended contents, the Côte de Nuits wines, were in such short supply thanks to the extremely difficult growing season, that the vats were used for Mercurey instead.

Thanks to the lighter touch in the winery the house style now relies on juicy fruit rather than heavy extraction to make an impression. The Gevrey-Chambertin 1er cru Petite Chapelle, which borders the *grand cru* Chapelle-Chambertin, is a good example of the new dispensation. It has the expected weight but this is given poise by tingling acidity, ripe fruit and crisp tannin.

Much has been achieved at Château Philippe le Hardi in recent years – made possible by being owned by the Crédit Agricole bank – but the journey has just begun. Never was the phrase 'watch this space' more apposite.

Try this: Aloxe-Corton Les Brunettes et Planchots

An essay could be written about the origins of this name which, in brief, derives from '*brunettes*' meaning dark or muddy soil and '*planchots*' referring to duck boards or wooden walkways to make traversing the ground easier. The 5-hectare vineyard lies below the village and looks up at the Hill of Corton, and the wine is neither muddy nor dark. The initial impression is of crisp and crunchy fruit, amplified by tingling acidity and carried into the finish by a stern note of tannin. Only then does a sweeter, more ample element make its presence felt.

Domaine Jean-Marc Vincent

3 rue Sainte Agathe, 21590 Santenay
Tel: +33 3 80 20 67 37

Few winemakers devote as much thought, followed by immense physical effort, to their viticulture as Jean-Marc Vincent. His constant examination and analysis of his philosophy and practices – tweaking here, adjusting there – translate into white wines with cut-glass flavours that yield slowly to the passage of time, gradually gaining flesh and breadth, so that by their tenth birthday youthful leanness has segued into mature succulence. The reds combine vigour and finesse, in that order, and are equally rewarding of a decade in the cellar.

Vincent was born in Lyon, grew up in Alsace, studied winemaking in Dijon and ended up in Santenay almost by accident when his maternal grandfather's vineyards became available to him unexpectedly. They

THE VILLAGES AND PRODUCERS OF THE CÔTE DE BEAUNE 253

were in poor condition, having been rented out to a less than diligent vigneron. Many vines were dead and one of Vincent's first tasks was the complete replanting of his holding in the *premier cru* Passetemps at a density of 14,000 vines per hectare. That was 1998 and in the years since he has continued to plant at higher densities than is the norm, most recently increasing to 17,000 in some plots. He also trains his vines higher (and has done so since before it became fashionable) which precludes the use of a vine-straddling *enjambeur* tractor, using a tracked *chenillard* instead. His vineyards are easy to spot, as in Puligny-Montrachet, where the higher vines form a 'notch' on the otherwise linear horizon formed by the neighbouring vines. More than one Côte d'Or winemaker has commented that if you want to see how to tend a vineyard, you should have a look at Vincent's.

His argument in favour of higher density planting is simple; the vines produce high quality grapes at a relatively young age, obviating the need to wait decades to get prized 'old vine' complexity. For the white wines the grapes are harvested into 20-kilogram boxes, crushed and pressed, given a quick settle and then transferred into 350- and 500-litre barrels for fermentation. There is minimal bâtonnage and after about 18 months the wine is racked into stainless steel. No sulphur is used in the winery and only a small amount at bottling. The red wines utilize between 50 and 70 per cent whole clusters, though the lower figure is likely to be the norm in future. Fermentation is in small stainless steel vats, allowing different terroirs, even within the same vineyard, to be handled separately. Vincent doesn't like a cool maceration as he aims for a light extraction, with a few pump-overs and few punch-downs. He neatly sums up his approach in the winery: 'The main idea is not to waste what we have done in the vineyard.'

Though Jean-Marc Vincent's name appears on the label this is very much a team effort, with his wife, Anne-Marie, taking care of all the administration, and his daughter, Anaïs, ready to join him late in 2025 after oenology studies and various stages with French winemakers, culminating in three months at Jonata in California.

Try this: Soler'Al

'Unique' is a tired and clichéd adjective, attached willy-nilly to any number of things that are far from one-offs. Its use to describe this wine is accurate, however. A hint to its vinification lies in the name, which references sherry's solera system. While not quite replicating that, the wine is blended across several vintages and given extended élevage in large oak.

254 CÔTE D'OR

The result is an intriguing combination of Aligoté's signature perk, the flavour twist that keeps the wine lively, along with an overlay of fuller, savoury elements that give breadth and depth. The result is an almost chewy texture and the flavour continues to develop long into the ample finish.

MARANGES

The Maranges appellation came into being in 1989 when the three sister appellations of Cheilly-lès-Maranges, Dezizes-lès-Maranges and Sampigny-lès-Maranges were amalgamated. For once, the bureaucrats got it right, their efforts yielding simplification rather than the more normal complication. Hyphens abandoned, Maranges set about creating an identity for itself – though the fact that administratively it sits in the *département* of Saône-et-Loire while vinously it belongs to the Côte d'Or has not helped. Many consumers consider Santenay as the Côte de Beaune's concluding commune. It's their loss.

In many respects, Maranges echoes the rise to recognition of Marsannay at the other end of the Côte d'Or. The best wines are delicious and are noted for good value too. The three villages are worth exploring, not for architectural splendour, but because they carry an echo of more rustic, less prosperous times in the côte. In Maranges you feel far removed from the manicured splendour of, for instance, Puligny-Montrachet less than 10 kilometres to the north. There's a wild, untamed feel, given emphasis by the topsy-turvy landscape, as the Côte d'Or peters out, not with a whimper but a dramatic flourish, curling to a halt like the business end of a hockey stick.

Producer

Domaine Chevrot et Fils

19 route de Couches, 71150 Cheilly-lès-Maranges
www.chevrot.fr; tel: +33 3 85 91 10 55

Pablo and Vincent Chevrot represent the third generation at this domaine of 20 hectares covering 15 appellations, established in its current form by their grandfather in the 1930s. Save for some holdings in Santenay this is a Maranges domaine through and through. The scale of the holdings obviates any need to buy in grapes, leaving the brothers completely in control – from planting the vineyards to bottling the wines. They replant about half a hectare at a time and always leave some land fallow to regenerate. Vine age averages about 30 years, with

THE VILLAGES AND PRODUCERS OF THE CÔTE DE BEAUNE 255

Pinot Noir making up 60 per cent of the vineyards, followed by 30 of Chardonnay and 10 of Aligoté.

Of those Aligoté vines one senses that Pablo is particularly proud of the parcel of 60-year-olds that occupy a single plot and which yield 'Tilleul', a wine that he describes as: 'Dry but not sharp, a little oily and salty.' It's an apt description of a wine with lovely grip and character, the citrus notes rounded by some smoothness in the texture. Moving to Chardonnay, the Maranges Blanc marches to a similar beat, augmented by some tropical fruit on the nose and a firm mineral snap on the palate. Taking things up a notch, the Maranges 1er cru La Fussière, fermented and aged in large oak foudres, comes with greater depth and more richness, balanced by lighter floral notes on the nose and a mineral, herbaceous whiff on the finish.

The vineyards have been certified organic since 2008 and a proportion are ploughed by the two horses the Chevrots own. La Fussière is such a steep vineyard, however, that the plough must be operated by a winch anchored at the top of the slope and moved laboriously from row to row. Fussière is also the source of a red wine that outstrips its sibling thanks to plump fruit and lovely succulence, leavened by a flinty, mineral bite. As in many parts of the Côte d'Or, the vineyards are confusingly named – the Chevrots produce both a Maranges 1er cru Les Clos Roussots and a Santenay 1er cru Clos Rousseau. Both have delicious, immediate appeal, the Roussots edging ahead thanks to a more complex flavour profile that adds some mild spice to the rich fruit.

About 50 per cent of production is exported globally, major markets being the USA, the UK and Australia, though that figure can go higher and is driven, or restricted, by economic conditions in the export markets. As Pablo Chevrot puts it: 'It depends on where the crisis is.'

Try this: Maranges 1er cru Le Croix Moines
French scholars may wonder why the masculine 'Le' precedes the feminine 'Croix' in this vineyard's name – and they will be left wondering, for there is no accepted explanation. Chevrot owns one-fifth of a hectare of the 1-hectare Le Croix, described by Pablo Chevrot as 'the jewel of Maranges', before he reflects, 'I wish we had more.' From that small plot the brothers make a fine and elegant wine, sweet red-fruited and far removed from the country cousin stereotype that still clings to Maranges, despite much recent evidence to scotch it. Age begets beauty here – eight to ten years is ideal.

6

RECENT VINTAGES – ENJOYING BURGUNDY

Generalizations and burgundy go together like oil and water, especially in the case of vintage assessments. Broad brush-stoke ratings – three stars for this vintage and four for that – carry a spurious authority and, except in the case of an unequivocally great or poor vintage, they obscure as much as they reveal. Côte d'Or burgundy is about nuance and shades of difference; it is celebrated for the remarkable diversity that can exist between adjacent vineyards, an attribute well recognized by commentators and consumers. The paradox is that those same commentators and consumers then ignore those cherished differences, seeking to paper over them by applying a 'one size fits all' approach to proclaiming the merits or otherwise of a particular vintage. You can't have it both ways: if you celebrate the remarkable diversity within the Côte d'Or then you must make allowance for that when appraising the vintages. Even the best vintages have dud wines and the worst can throw up pleasant surprises.

This is best illustrated by the run of vintages since 2010, many of which were beset by myriad difficulties, most particularly problems of spring frost and summer hail. These are the twin terrors facing *les vignerons* today. Frost and hail strike with such apocalyptic force that little can be done to mitigate the destruction unlike, for instance, an outbreak of mildew, which can be fought against with some success, however enervating the battle might be.

I would not want to trivialize the hardships faced by small growers when hail or frost hits but the sweeping reports that usually follow a

Guarding against frost, Chevalier-Montrachet 1 April 2023: 'tow and blow' trailer mounted fan to circulate cold air

hailstorm in the Côte d'Or, suggesting total wipe out of a year's crop, are no use to anybody. They might make good headlines but that is all. And when hail strikes, the côte's bewildering fragmentation of vineyard ownership ameliorates the impact; because a grower's holdings may be spread over half a dozen appellations it is highly unusual for all to be hit at once. It's small comfort but it can make the difference between producing almost no wine to turning out a well below average, though still viable, quantity.

Notwithstanding all of the above, we cannot entirely escape generalizations when offering vintage assessments in the Côte d'Or. Without them we would simply get bogged down in qualification after qualification. That said, it is most important to stress that the assessments that follow are not carved in stone. I believe they are valid at time of writing but if I were to revisit them in five years' time there would be a host of changes, based on further evolvement in the wines: decline in some, blossom in others.

All vintages from 1985 to 2023 are assessed and eight historic vintages of high repute are noted for interest. Extreme caution should be exercised when considering a purchase of old wines. How they have been handled and stored will affect their quality. If they have been on

258 CÔTE D'OR

the auction merry-go-round they might have been subject to all sorts of deleterious influences, most particularly rapidly fluctuating temperatures. And then there is the matter of fraud, though this really only applies to a handful of banner names whose wines regularly fetch high prices at auction.

It is possible to strike it lucky and unlucky. In my case, luck came by way of a small parcel of old Ramonet *premier cru* Chassagne-Montrachet *rouge* from a number of vintages, including 1966 and 1969. All the wines were good but the Clos Saint-Jean 1969, drunk in 2009 to commemorate the fortieth anniversary of the first moon landing, was superlative, with lively acid, good fruit and wonderful length. The final bottle was drunk a decade later and it was still marvellous. Ditto a Domaine Jessiaume, Santenay Gravières 1929, drunk in 2020. It came directly from the domaine's cellars and, if a little shrill, was still eminently drinkable. It is wines like this that tend to get talked about and remembered; the duds are air-brushed from memory, creating the impression that old treasures abound. That is not the case, as illustrated by an Armand Rousseau Chambertin 1953 tasted in 2016 that was a dull, lifeless brown in the glass with a sour nose and a brackish palate. Caveat emptor.

VINTAGE REPUTATION

Initial vintage assessments quickly get written in stone. The trade and press critique the wines, write their verdicts and in time move on to the next vintage. Only the most diligent revisit and re-taste as the wines mature, to provide running commentary rather than fixed snapshots. Aside from those assessments and ratings, delivered when a vintage is released, it is worth remembering two other influences on its reputation. The first is partially reliable. The second is useless yet is remarkably persistent in consumers' minds. In the first instance, a vintage's reputation is primarily set by the quality of the red wines; 2005 and 2009 are good examples of this. When one pithy comment is required to rate a vintage it is the quality of the red wines that will see it lauded or dismissed. And in the second, the reputation of the vintage in Bordeaux, hundreds of miles across France with a different climate and grapes, can greatly affect the worldwide impression of the Burgundy vintage. A good example of this is 2013 – generally dismal in Bordeaux but tasty in Burgundy. Beware of instant pigeonholing based on little concrete evidence.

(2024)

Only a preliminary assessment is possible for this vintage, recently fermented and still in barrel at time of writing. 'It hasn't stopped raining since last October', was a constant refrain from many of the dozens of winemakers I visited in July 2024. Some described it as the most challenging season in a generation. The rain led to serious problems, especially mildew, and a much reduced crop, especially in the Côte de Nuits; yields in the Côte de Beaune were much more encouraging. In its own way, 2024 was as challenging as 2021. Early indicators are that it might yet deliver similarly attractive wines.

2023

An early September heatwave that was widely welcomed at first brought a much-needed boost to ripeness levels, but after a week it threatened to generate a surge in alcohol levels in the large crop. Harvesting rapidly, with some domaines starting at 6.30 a.m. to avoid the worst of the heat, was the secret, but this then delivered a big challenge in the cellars in the form of a huge quantity of grapes that had to be handled quickly. The vintage was a success in both colours, if a lot more challenging than 2022. The wines have easy appeal, though it is too soon to cast definitive judgement. Considering the much reduced quantities in 2024 it is worth taking a close look at 2023.

2022

Destined for greatness, 2022 combines the richness of 2020 with the reserve of 2021. Frost threatened in early April but didn't strike with anything like the venom of 2016 or 2021. Heavy rain in June proved ultimately to be a blessing, for it was hot, dry and sunny thereafter. Harvest was in full swing in the last week of August. The white wines are very good, verging on excellent, majoring on purity, freshness and finesse. This is a vintage to explore the Bourgogne Côte d'Or appellation from good growers: the wines punch well above their weight. The reds are marginally better, approachable young thanks to bags of sweet fruit. But they are far from being easy charmers; there is depth and substance to back up the immediate appeal. Vintage 2022 is a developing classic, a winner in both colours, particularly the reds.

260 CÔTE D'OR

2021

Savage April frost reduced the crop considerably, yet small quantities of attractive wines were made. After three hot, ripe vintages 2021 is an outlier, a step back in time to a period when rich abundance was not a given, when attaining full ripeness was a struggle. The flavour register is lighter, the wines are charming rather than profound, and to fully appreciate them some palate 're-calibration' is needed. The ample flavours of the previous trio are absent in 2021, replaced by a lean, racier palate profile. The wines display a lighter weave in direct contrast to the denser textures that had become commonplace and which risked warping the grace and elegance for which Burgundy is renowned.

2020

The Covid vintage. Despite lockdowns and restrictions it went well, though it is difficult to describe the pall of uncertainty that hung over every aspect of the work of *les vignerons*, from vineyard to cellar. An early harvest, with picking well underway by 21 August in hot, though not torrid, conditions. In time, the white wines may come to rival 2017 and 2014, mainly thanks to a delicious streak of acidity. They are already drinking beautifully, with excellent interplay between tropical and citrus notes. The reds are less certain, a hearty flavour register characterized some, thanks to baked fruit flavours and high alcohol. That said, the best wines are balanced and should reward long ageing.

2019

Brain-bending, week-long heatwaves in June and July, when the mercury regularly touched 40°C, raised concerns that the wines would be 'branded' by the torrid conditions – the reds baked and jammy, the whites lush and lumpen. Mercifully, this has not proved to be the case – despite the scorching summer, remarkable freshness was maintained in the wines. Acidity was the key to quality in both colours. The whites have great energy with penetrating fruit flavours, rich rather than lush. The reds have concentrated fruit – not oppressive thanks to freshness from the acidity. Traditionally 'lesser' appellations such as Marsannay and Maranges gained by way of appealing ripeness.

2018

Lots of winter rain and oodles of summer sun is 2018 in *précis*. The result was a large, ripe harvest that was fully underway by the last week in August. The challenge was to harness the richness and, in general, this was better achieved in the reds than the whites. Many whites are imposing on the palate, more impressive than gracious. The reds run to lavish flavours, mouthfilling and in some cases a little brusque. A hearty flavour profile will divide opinion; it is a question of quality versus style. The former cannot really be faulted, the latter will delight those who like sumptuous reds but not those who seek finesse and elegance in their burgundy.

2017

An outstanding year for white wines and – so far, with fingers tightly crossed – this writer has yet to experience a premoxed bottle. For once, the whites usurp the reds' place in the limelight courtesy of remarkable balance, freshness and persistence. Harmony and elegance abound. Comparisons with the other stellar white vintage from this decade, 2014, will run for many years. If the whites are spectacular the reds are safe and sound, and will provide satisfactory drinking in the short to medium term.

2016

This was a year of meteorological infamy. The night of 26/27 April will long be remembered in the Côte d'Or for the devastation wrought by frost. This was followed by a wet May leading to problems with mildew in June. The crop was meagre – from some vineyards no wine at all was made. Yet some excellent wines were produced in both colours, though they are already scarce and hard to find.

2015

An early harvest, the first pickers in Meursault were in the vineyards on 27 August. The grapes delivered concentrated juice, the small size of the crop being the only concern. The challenge was to avoid too much richness, particularly in the whites, and it was better met this year than in 2009. Getting the harvest date correct was critical. This year promises to develop into a stellar year for red wines. The fruit is ripe, there's

262 CÔTE D'OR

structure and body too, all components are present in balanced measure. Some wines may be a little flashy but that is a minor concern.

2014

After a challenging season the prospect of a good harvest looked remote until September sunshine came to the rescue. The white wines are superb: vital, intense and structured. There's measured richness without lushness. At a decade old the *premiers crus* are excellent, with the *grands crus* still improving. In the reds a pleasant austerity contrasts markedly with the flesh of 2015; the wines are upright and correct rather than charming. The 2014 vintage is destined to be overshadowed by 2015, just as 2008 is by 2009, making it a vintage for wine drinkers rather than trophy-hunter collectors.

2013

The vignerons struggled all through a very mixed growing season, yet in the end 2013 turned out to be a year of enjoyable surprises in white and red. The whites show good fruit, pleasant and clean. The reds are light-bodied, though late-picked examples can be coarse.

2012

The weather gods presented many challenges in 2012, though the upside was that the problems thus caused reduced the yield rather than the quality of the small crop that eventually resulted. The whites are classic, poised and elegant, balanced and intense, with a mineral backbone that should guarantee long life. The reds possess abundant fruit, rich and balanced, with ripe tannins adding backbone. Born out of difficult conditions that are not reflected in the wines, the best will age well while the lesser wines are drinking nicely now.

2011

Only the diligent growers managed to wrest appealing flavours from the grapes that survived a roller coaster growing season, and they produced tasty wines, full of charm and bright fruit if not great weight or depth. That said, there is variation in the wines; some are thin and unremarkable others have purity without power. The whites are fresh and easy, the reds have nice red berry fruit with light tannins.

2010

The rush to judgement tripped up many commentators this year, writing off the vintage before a grape was picked. Small bunches led to low yields but what wines were made, both red and white, are delicious. The whites are fresh and clean with mineral intensity, rivals to 2008 and certainly better than 2009. The reds have energy and vitality, and more reserve than 2009. Precisely flavoured, with good acidity and a touch of pleasant austerity, they promise a long life.

2009

This was a highly rated vintage from the outset and was released with much hype and trumpeting of its virtues. It is a red wine vintage. The whites are not as successful, for the conditions that yielded such impressive flavours in the reds endowed many whites with an overly rich, plump character. As ever, there are exceptions but too many of them are plodding on the palate. The reds are mouthfilling and sumptuous. The wines are more expressive of the vintage than their terroir and whether they are equipped for a long life remains to be seen.

2008

Much vineyard work and strict selection was needed to harness the potential of a small crop. Acidity is the key to 2008. The whites have good, precise flavours and appealing fruit and freshness. The reds are forthright, singular and intense, lithe but not lean wines. They look destined to be defined by what they are not: not 2009, not opulent, not lavish. They are racy and fresh and, though overshadowed by 2009, 2008 may outrun it eventually.

2007

A mixed vintage, more consistent for whites than reds. There was good sunshine in early July but some storms and heavy rain thereafter. Much of August was grey and wet but the sun returned in September, together with a drying breeze. The reds were awkward in youth, marked by promise on the nose and mean flavours on the palate. They are still variable, but some have softened to reveal pleasant fruit and real attraction.

2006

Spring was late after a long-drawn-out winter. July was hot and dry, August cool and wet. Harvest took place in the second half of September. On early tasting the whites were impressive but age has brought on plumpness and weight. The reds have reasonable structure and elegance without appreciable depth. Some are on the lean side. In general, there is better texture and richness in the Côte de Nuits than the Côte de Beaune.

2005

The first 'vintage of the century' of the noughties. Winter was cold and a wet spring followed. Flowering was slow and there was significant *millerandage*, which favoured concentration in the reds. It is the whites' lot always to be considered an afterthought to the reds, yet they are by no means poor. They are full-bodied, running to lush at times. The reds have density and depth without being overly heavy thanks to excellent balance and great vitality. From first sniff to last sip they exhibit a certain gravitas that marks them out as serious keepers that will mature and evolve for decades. Approaching their twentieth birthday some of the reds are still closed and forceful; more patience than initially supposed will be needed to enjoy these wines at peak maturity.

2004

Variable weather from beginning to end was the hallmark of 2004. The result was a mixed vintage, the charms or otherwise of which have been the subject of much debate and comment. The whites were marginally better than the reds, pleasant in the short term with reasonable freshness. The reds were marked by an unpleasant barb of flavour, which has softened with the years but only the top wines are still worthy of interest.

2003

This heatwave vintage will divide opinion for as long as there are bottles to drink. Daytime temperatures regularly hit 40°C in August, with little remission at night. I have never liked these furnace-formed wines. Ill-defined flavours abound in the whites; the fruit is lush and lavish, almost oily. The reds have a baked quality like overcooked jam. Many of them taste like Pinot Noir grown in the Rhône Valley. Yet they have their advocates, who counsel patience. Perhaps 1947 tasted similar in youth. Perhaps a half-century in a cool cellar will tame the torrid flavours. Perhaps.

2002

The whites have a racy purity in marked contrast to 2003; there's good fruit and balance, and for sheer style they are marginally better than their red siblings. The reds are fine flavoured without too much weight, possessed of lovely fruit and smooth tannins. This has always been a favourite vintage for both colours, though some reds are now beginning to dry out a little.

2001

A mixed year led to uneven ripening, which called for diligent sorting and those that did made reasonable wines. The whites were marginally better than the reds with good acidity in a light framework. The reds had fresh fruit in a delicate structure that lacked robustness.

2000

The 2000 vintage will always be unfavourably compared with 1999, resulting in it never establishing a clear identity of its own. The growing season was mixed, with a noticeably cool July. The whites had good fruit and an appealing character without great concentration and were a little better than the reds. These were varied and needed to be assessed on a producer by producer basis, which has made it difficult to form a general impression of their virtues or otherwise.

1999

Will always be remembered as a red wine year but the whites were of good quality too. June and July were mixed but August was hot and dry, and sunshine in September brought the grapes to good ripeness for harvest to start in the middle of the month. The wines had all the components in good measure, led by rich fruit, though I have come across some where the tannins were brusque and a little dominant. Once these soften the wines are delicious. The best are now glorious, wondrous examples of burgundy at its finest. All are mature but there is no hurry with the grander wines.

266 CÔTE D'OR

1998

A tricky vintage where just about every difficulty was thrown at the growers, including spring frost and summer hail. Yet some surprisingly attractive wines were produced, safe and satisfying rather than spectacular. Good fruit rather than backbone characterized the whites. The reds could be a little mean but once the tannins lost their edge there were plenty of charming wines.

1997

The weather conditions during the summer hardly formed a pattern, alternating between hot and cold. September brought welcome sunshine and harvest began in the middle of the month. This was a white wine vintage and they ran to sumptuous and plump, with mouthfilling flavours. The reds were more variable, ranging from some delicious wines to others that were hollow and a little charmless.

1996

A favourite red wine vintage. June was sunny, July and August less so, and September sunny again. The crop was healthy and large. The whites were enthusiastically received and started out life charged with vitality and no lushness, until the scourge of premature oxidation hit. For the reds, 1996 has regularly attracted criticism for being overly acidic yet I have always enjoyed the wines and have only occasionally come across sinewy examples that could have done with more flesh.

1995

The weather was lovely in July and August but September was gloomy and intermittently rainy. A small, high-quality vintage resulted, producing some lovely wines in both colours. The whites were fuller and richer than the 1996s but with good acidity. Sorting out of grapes that had not reached full maturity was the secret to making good reds. The best had ample fruit and good concentration and the top wines are still drinking well.

1994

Rain before and during the late September harvest dashed hopes of a fine vintage. As a consequence the results were generally limp, though not without some charm. The whites made for easy, early drinking. The reds had an eviscerated quality, delivering some decent flavour but not persisting on the finish.

1993

A vintage that enjoys a mixed reputation, rated highly by some, though never accorded universal acclaim. This was a red wine vintage, the whites were variable. The calling card of the reds has been impressive acidity, the wines are intense, not fat, and boast a slight rusticity that divides opinion.

1992

A white vintage. The summer weather was unremarkable, free of storms, and good sunshine preceded the harvest which began in mid-September. The whites had lovely fruit and succulence with moderate acidity, and a well-stored example should still offer satisfaction if not excitement. The reds drank pleasantly in their first decade and never promised anything greater in the second.

1991

Coming after the previous trio, 1991 always struggled to establish an identity for itself. Frost in April and summer hail reduced the crop. Good sunshine preceded the harvest towards the end of September. The whites were serviceable if not distinguished, the reds were better.

1990

A great vintage, especially for reds. A hot summer followed a mild winter, leaving the vineyards badly in need of the rain that fell in August. The whites have always lived in the reds' shadow, and they showed less consistency, but the best were marvellous and still should be, glorious examples of pre-premox white burgundy. The reds were rich and robust from the outset, the concentration amplified by *millerandage*. Because of the summer heat, 1990 was marked by the vintage rather than the terroir and the dense, full-flavoured wines are now magnificently mature.

1989

Two hundred years after the storming of the Bastille, the Côte d'Or celebrated with an excellent red wine vintage and a good one for whites. Winter and spring were mild, summer hot and sunny, leading to a relatively early harvest in mid-September. The whites boasted lush fruit

268   CÔTE D'OR

with lowish acidity. In style the reds lay somewhere between the lavish 1990 and the crisper 1988 and drank well from an early age. Only the best are worth seeking out now.

1988

An opinion divider. A mild spring and a hot summer, the depredations of frost and hail missing for once. It was a large crop for the whites, perhaps too large, for a lack of body was a frequent shortcoming. The reds were generally well regarded though they had their detractors who found them acidic to the point of unyielding angularity. There were some mean wines and others that, as they aged, showed too much sinew and not enough flesh, but this couldn't detract from the quality of the majority.

1987

Of little interest today. A difficult growing season led to a late harvest in October. The whites were variable, the reds more serviceable.

1986

Lauded for whites more than reds. Many of the whites drank well in their first decade but failed to stay the pace after that. The reds attracted some hoopla initially but subsequently failed to live up to the hype. Variable quality was the big problem; the good wines were quite attractive.

1985

A bitter January saw the côte gripped by arctic temperatures that killed vines in places where the temperature settled below -20°C for a period. Mother Nature was more benign after that: summer was reasonable and there was good sunshine in September, leading to harvest towards the end of the month. The whites were mixed and never caught the imagination the way the reds did. These were delicious. They relied on lovely fruit and good balance to make an impression, not overt power or weight. Today, only the best remain vibrant, others are heading down the curiosity road, of interest to those who like the flavour quirks of old age.

Older vintages of note

The four decades prior to 1985 are notable for the preponderance of dismal vintages, almost an inversion of the picture since. The years 1974, 1968, 1965, 1963, 1958, 1956 and 1951 constitute a roll-call of misery for producers and consumers. Nonetheless, there were some excellent years, eight of which are profiled here.

1978

Rescued by late sunshine. The growing season was mixed and prospects were poor until the weather turned good in August, leading to a small but excellent crop in October. This vintage is the standard bearer for a too-often mediocre decade.

1969

The best wines are fading now but 1969 produced reds of great charm; the flesh has fallen away but the spine of acidity is still vibrant. Good examples are now frail and delicate but still tingling.

1964

A brutal winter was followed by a searing summer, resulting in wines of concentration and depth, with good balance too, attributes that endowed them with impressive ageing ability.

1961

A very good vintage in Burgundy though not as celebrated as in Bordeaux. Rivalled by 1962. Of curiosity value now.

1959

An excellent vintage, yielding wines capable of ageing for half a century. The growing season was near to ideal giving a bountiful, high quality crop. Right through their life the wines have been impressive and it is still possible to come across vibrant examples.

1949

Vies with 1947 and 1945 as the best vintage of the 1940s and probably just shades it. A variable spring preceded an excellent summer; the crop was smallish but the quality was high.

270 CÔTE D'OR

1947

A superb summer led to an early harvest, the only problem being the risk of high fermentation temperatures and consequent baked flavours in the wines. Those that avoided this pitfall made wines of legendary quality and longevity.

1945

Vineyards across the continent appeared to celebrate the end of the Second World War by producing remarkably good wines, albeit in small quantities. Burgundy was no exception.

ENJOYING BURGUNDY

The Burgundians enjoy their wines with notably less ceremony than attends the broaching of a grand old bordeaux, so a leaf taken from their book is the best starting point when considering details of service and presentation. Beware over-sanctification and undue ceremony – they get in the way of enjoyment – but be prepared to make an effort as a good burgundy warrants thorough assessment. Above all, pay attention to the wine, don't treat it like background music, put it centre stage to be tasted receptively. 'Listen' to the wine.

Nerdy accumulation of detail and facts relating to viticulture and winemaking may help in appreciating wine. They also may not. They may obscure the wine itself in a 'wood for the trees' way. It is like tasting a great dish – *boeuf Bourguignon* comes to mind – and then trying to describe it by identifying and listing all the ingredients. We don't rate a car by listing its components, so why do it with wine? It is the combination and interplay between all the diverse elements and influences that make wine, especially burgundy, endlessly fascinating. We should pay attention to that, the interplay and harmony, the overall picture, and not the paint used in it. Simply making a list of scents and flavours is wine tasting by numbers; it may demonstrate knowledge but hardly understanding or appreciation, least of all enjoyment.

Today's fad for the grand tasting where dozens of venerable treasures are opened for assessment (though hardly enjoyment if they are all to be spat) holds little appeal. Could a dozen of Shakespeare's plays or Mozart's symphonies be assessed by studying only the opening passages of each? I think not. Far better to open half a dozen bottles for careful scrutiny and then enjoy them over a meal with company.

Serving

To serve a white burgundy too cold and a red one too warm is a vinous crime. In the first instance the wine's flavour is stunted; it has been whacked on the head and is dazed and uncommunicative. In the second, the wine loses definition and focus, the flavour becomes soupy and the signature Pinot Noir tingle, the electric charge of flavour that runs through all decent burgundy, is lost. White that is too cold will warm up, red that is too warm can be helped by a few minutes in an ice bucket but getting it right in the first place is preferable. A range of about 10 to 12°C works well for whites. As a guide for serving reds, 'room temperature' has had its day; it is perhaps fifty years out of date and harks back to a pre-central heating age when homes were much cooler than they are now. About 15 to 17°C is correct, and if it starts a little too cool it will warm in the glass. The bottle should feel cool to the touch.

Whether to decant burgundy or not is as much a matter of personal preference as anything else. The Burgundians seldom do it. If it is a young wine that needs some air, red or white, then it should be splashed into the decanter and not poured gingerly down the side. Serving a fragile old red at a dinner, I may not decant it but will assemble all the glasses on a side table to pour it almost in one go and then distribute them rather than going round the table with the bottle. The green glass in these bottles is usually much paler than that used today so it is easy to monitor the advance of the sediment along the side of the bottle as you get to the end of the wine. Having the glasses close together eliminates all but the gentlest back and forth movement that stirs up the sediment.

With a wine like this, I serve it on its own after the starter and then segue into the main course and recharge the glasses with something firmer. Keeping such a treasure for the end of the meal is not a good idea. It is far better to serve venerable wines early on, not first but early on, and then follow them with younger, more robust ones. Serving from old to young may not seem logical but a younger and fresher wine will deal with the inevitable palate fatigue better than a frail old treasure. The more vigorous wines can stand coming last much better than the pensioners.

272 CÔTE D'OR

Glassware

From a time not long ago when the Paris goblet was ubiquitous and served for almost every style of wine, the world is now awash with an ever expanding and bewildering range of glasses from which to enjoy one's wine. It is only a slight exaggeration to say that there is now a glass for every wine. Finely crafted and exotically shaped, many possess a sculptural quality that makes them objects of beauty in themselves. Some, however, are designed to the point of distortion, sacrificing aesthetics in pursuit of efficacy. There is no need to trade beauty for ugliness; form and function can co-exist without conflict. Fortunes can be expended on them, yet fortunes do not need to be spent – a general purpose glass for white and another for red is enough to start with.

Noting which make of glass was used by about 50 of the domaines visited for this book showed two clear 'leaders': Zalto was favoured by 15 of them and Spiegelau by 12. Next came Lehmann with five, followed by a wide range of others with a small number each: Riedel, Chef & Sommelier, Sydonios, Grassl, Gabriel Glas and Royal Glass. Riedel does not enjoy the same popularity in Burgundy as it does elsewhere, though it shouldn't be forgotten that Spiegelau is owned by Riedel. Though unscientific, what this survey indicates is that consumers should not get too hung up on choice of glassware. Every glassmaker will tell you otherwise but their 'counsel' is designed to sell, not impart independent advice.

What nearly all of these glassware ranges have in common are ultra-fine stems and wafer-thin bowls, making them fragile yet excellent vessels for showcasing and enjoying the Côte d'Or's finest wines. Handle with care but don't forget that the majority are dishwasher-friendly and that you are more likely to break one when hand washing than when consigning it to the machine.

Food and wine

The old adage 'white with fish, red with meat' has taken a bashing in recent years, with writers on the subject falling over themselves to debunk it as a relic from a less sophisticated age. Yet as a tenet to fall back on when choosing what wine to drink with what food it has not been bettered. 'Drink what *you* like,' is a noisome piece of advice and abrogates responsibility on the part of the person offering it. If its aim is to banish inhibition and intimidation then it is perhaps laudable but it is otherwise useless. The polar opposite is the position occupied by the high priests of food and wine matching who treat it like a pseudo-science, supposedly winkling out a DNA match between nectar in the glass and ambrosia on the plate. A common-sense middle ground serves best.

Do pay attention to food and wine matching and explore beyond the old tenet but don't throw the baby out with the bath water, because there's still wisdom in it. Done properly, a good food and wine match has a hand in glove inevitability to it, each enhances the other, each flavour is brought into focus by the other, and the enjoyment of the occasion is greatly heightened as a result. Some matches do seem to have come down on tablets of stone: *ris de veau*, a staple on many Côte d'Or restaurant menus, paired with a broad-bellied Meursault is sublime, a rich and satisfying juxtaposition of patrician flavours; *boeuf Bourguignon*, long simmered to richness, needs a great Gevrey-Chambertin, and if you can manage it, Chambertin itself, the meaty depth in each finding an echo in the other; *poulet de Bresse*, which, depending on cooking method, pairs equally well with white or red, should be matched with a wine that boasts elegance rather than weight.

Traditional Burgundian cuisine is hearty, best described as 'full not fine'. Of late, restaurants are offering less rich alternatives that chime well with the less oaky, less extracted, more transparently flavoured wines being made today. A good rule of thumb is to try and match the weight and depth of flavour in the wine to that in the dish. I hesitate to be dogmatic but I will offer one diktat: époisses cheese, one of my favourites, can cut the legs from under the greatest of wines, indeed the greater the wine the greater the crime in serving it thus. It's like throwing it into a bear pit. A rich white might just cope, though if I was to suggest anything, experience has shown top notch Alsatian Pinot Gris at a decade old to be best.

Ageing

Forecasting how a burgundy will age makes predicting the adult an infant will become look like an exact science. Sadly, it is a question that gets little asked today. Outside of a small coterie of wine lovers considerations of how a burgundy will age are seldom addressed; it is a subject of great interest to a diminishing number of people. Across the whole wine world, wine is being made to taste good *now*. Speculating on how

274 CÔTE D'OR

it might develop and hopefully improve is increasingly the province of the wine anorak. Yet the potential for change, development and improvement is what sets fine wine apart from everyday wine. The mature whole is greater than the sum of its youthful parts. Time unlocks the potential, making the wine more complex, more satisfying, and its passage eventually leads to decline, decay and corruption. Time acts like an emulsifying agent, melding contrary components into harmony: yielding a truffled fruit character in the whites, sweet incense in the reds. Catching a wine at the right moment is the key. It's a matter of taste but as a general rule, for whites and reds, I suggest keeping *village* wines up to five years, *premiers crus* up to ten and *grands crus* for ten or more years. Personally, I keep all categories for longer than this.

7

THE CÔTE D'OR TODAY

It is 80 years since the end of the Second World War, a period in the Côte d'Or's story that can be divided equally into the years before 1985 and the years after. Compressing those decades into a one-sentence summary: In the first half there was a handful of great vintages and in the second there was a handful of poor ones. It could be argued plausibly that a golden age began for the côte in 1985. The forty years since have seen fundamental developments in viticulture and winemaking, and resultant changes in the wines, that were nearly all for the good. The business model changed too: the structure of the wine trade, who made it, who bottled it and who marketed it, all changed radically. A broad brush-stroke picture of the trade for most of the twentieth century saw a clear divide between the domaines and the négociants. Though some domaines had been bottling their wine since the 1920s, the domaine-négociant model, where the first made wine and sold it to the second, to be assembled into what were essentially brands such as 'Volnay' or 'Gevrey-Chambertin', still held sway.

The distinction between domaine and négociant has become increasingly blurred in recent years, with each occupying some of the ground traditionally held by the other, to the point where the two are now overlapped and interlocked like never before. Most domaines now bottle and sell their own wines and a good number have entered the grape market as buyers to supplement their own fruit – the current land prices precluding expansion by vineyard purchase. The vineyard owner with grapes to sell occupies a prized position, able to command a hefty price, getting ready cash and having no further worries about vinification, bottling, marketing, shipping and so forth. A web of deals facilitated by *courtiers*, brokers who act as middlemen and get a percentage for their efforts, lies just below the surface in the côte, not deliberately hidden but obscured by its own labyrinthine complexity.

As the domaines went in search of grapes, so the négociants bought up whatever land they could in an effort to secure their supply and exercise greater control over the winemaking process from vineyard to bottle. Domaines with landholdings in prime vineyards became increasingly wealthy, on paper at least, while négociants with little or no vineyards were left high and dry as their supplies of top quality grapes disappeared and they had to settle increasingly for fruit of lesser potential from lowlier vineyard sites.

In parallel with the shifting structure of the trade went a marked increase in quality, prompted by a desire to make better wine, stimulated by competition from the new world and facilitated by the weather gods smiling on the côte like never before. Quality wise, there has not been a truly awful vintage since 1985. There have been challenging ones, such as 2016, 2021 and 2024, but it was quantity rather than quality that was badly affected in those vintages. The use of herbicides, pesticides and other treatments was reduced, a big yield was no longer seen as desirable and a piece of equipment now considered standard – the sorting table – became widespread. Above all, the vignerons' greatest asset, the golden slope, began to be cherished after decades of being thrashed with ill-considered applications of weedkillers and fertilizers. For people who believe so passionately in terroir they treated the land pretty shabbily in years gone by. A golden period, yes, but not without its challenges.

A PAIR OF CHALLENGES

Burgundy was hit by a perfect storm in 1996, though its catastrophic consequences only manifested themselves around five or six years later. Much has since been written and spoken about premature oxidation (premox for short and pronounced prem-ox, not pre-mox) and many, many bottles of white burgundy have been poured away because of its baleful effects. Many others still lie in the cellars of wine lovers and restaurants around the globe, though not every one of them is a ticking bomb. The random nature of premox is such that if two bottles are chosen from the same case one could be glorious and the other undrinkable.

No great ability or experience as a taster is needed to tell that a bottle is sub-standard. The easiest and most foolproof way to spot a corrupted bottle is as suggested above, two from the same case, one good, one bad. Supposing the wines are between five and ten years old, the good bottle will sport a bright colour and a vital flavour; you will want another sip. The corrupted wine will be its antithesis: noticeably dark for its age; no fresh lift of fruit aromas on the nose but rather a heavy and sweet smell like marzipan or cheap perfume; a lack of vigour on the palate, flat and featureless, no juicy fruit, instead a ponderous procession of flavours that veer towards baked fruit and oily nuts, with a sour twist on the finish. The least experienced taster can spot the difference between good and faulty.

Today, to be reasonably sure of getting a good bottle of aged white burgundy you need to reach back beyond the mid-1990s, so questions of storage, provenance and the wine's ability to age for that long in the first place come into play. All being well, the rewards are sublime: soft but full-flavoured, mellow fruit with nuts and spice and truffles and perhaps a sweet hint of decay, pleasant, not offensive.

When the problem first came to light in the early years of this century, bafflement and denial were the first reactions from producers and, given the time lag of about five years before the wines deteriorated, they had plenty of wriggle room, allowing them to suggest poor storage as the likely cause. The wines, after all, had left the winery in perfect condition and if the buyer had drunk them on receipt there would have been no problem. But white burgundy should age gloriously, developing and improving over fifteen or twenty years. It was only when bottles from cellar stock that had never left the domaine and which had been correctly stored were tasted and found to be faulty that the argument was found out.

On release, 1996 got the thumbs up and all was well until the early years of this century when its reputation began to unravel. There was no collective, industry-wide negative judgement, just a mounting series of questions and comments, as anxious consumers and concerned merchants broached a second or a third bottle, having found the first one from a long-anticipated case to be dismal. Initially, a faulty bottle could be dismissed as a one-off aberration, especially if a second bottle was found to be excellent. And consumers who decided to drink the 1996s shortly after purchase revelled in their quality, but those who acknowledged that quality by cellaring the wines for five or more years were in

278 CÔTE D'OR

for a shock. Many have now taken to drinking their white burgundies young, and they are delicious so there is no way of knowing if they will fall off the cliff a few years down the line.

Today we may well be out of the woods but it is difficult to know if the causes being suggested are correct, and if the solutions being implemented will be effective. This multiplicity of causes, allied to the time lag before the problem manifests itself, militates against the speedy implementation of a solution. At least it is now the case that most growers will talk openly about the problem and their sincere efforts to remedy it. It must have caused the conscientious amongst them great anxiety but at least they were paid for the wines; the consumer was the one out of pocket. There have been many explanations, not so many apologies and few refunds.

The 'perfect storm' nature of premox reveals itself when a search for the cause or causes is undertaken. First under the microscope were the corks used to seal the bottles. Because of burgeoning demand 30 years ago cork quality was at best variable, causing many Burgundian producers to switch to DIAM instead. In addition, competition from the New World was on the rise, prompting a move in Burgundy to produce wines that could be drunk young. This coincided with a switch from hydraulic to pneumatic presses that produced clean and pure juice, which was then transmuted into clean and pure wines. But the wines were fragile and lacked the 'stuffing' to protect them against subsequent oxidation. All beauty and no brawn. They were like child prodigies, brilliantly accomplished in youth, but bereft of the wherewithal to carry their class into maturity.

Another trend that was beginning to grip the global wine industry in the mid-1990s was the movement towards using less sulphur, in the form of sulphur dioxide, both during production and at bottling. Sulphur is an anti-oxidant and essentially acts as a preservative, stabilizing the wine against spoilage once bottled. Sulphur levels were reduced, in effect removing a safeguard, thus increasing the risk of oxidation. Excessive handling of the wine during élevage, principally through bâtonnage or lees stirring, has also been suggested as a cause of premox.

If it is ever proved beyond doubt that cork was the cause of premox then it is not fanciful to suggest that screwcap was the indirect solution. Thanks to its taking business from traditional cork producers it forced them to up their game and apply modern quality control practices to a slack production process. Thus the quality of corks today is superior to what it was 30 years ago. Save for a few instances, screwcap has not replaced cork; where cork has been abandoned by winemakers they have generally favoured DIAM, a 'cork' made from ground particles of cork compressed and bound by resin. The paradox is that the cork industry has been saved by its competition; if cork was the culprit in premox it probably no longer is.

It may not have been, however, and thus conscientious winemakers are examining vinification methods today like never before. Flawless juice is no longer so venerated and sulphur has shed its bad image. The white burgundies produced in 2014 are superbly good and so far I have not encountered corrupted bottles, ditto 2017, so fingers crossed. But whether there is any need to produce wines capable of ageing is worth considering. Changing tastes and drinking patterns indicate that, apart from a small number of consumers, there is little desire to cellar wine for an extended period. Premox may have urged people further along the road towards instant gratification and away from deferred enjoyment. As a wine style, aged white burgundy may have had its day. Future historians may write that the early years of the twenty-first century witnessed the disappearance of this wonderful wine style. If so, that would be a terrible shame.

Premature oxidation was, and is, the problem or challenge most likely to have affected consumer perceptions of the wines in recent years, though from the winemakers' standpoint the increasing threat of devastating weather events, in the form of spring frost and summer hail, was paramount. It is difficult to know which is worse. Both are extraordinarily destructive and invite grim-reaper comparisons for the viciousness with which they strike. In spring the temperature need only drop a couple of degrees below freezing to kill the delicate young buds. April is the month when they are at their most vulnerable but the vignerons don't sleep easy until mid-May. In an effort to beat back the elements they employ a range of measures to protect the vines: smudge pots known as *bougies*, fans to circulate the cold air (see photo, p. 257), fans used in conjunction with heaters to spread warm air and even helicopters that hover over the vineyards to stir up the air. As with hail, frost damage can be remarkably localized. A small declivity may encourage cold air to settle where an open slope allows the air to move more freely.

Bougies (smudge pots) are lit in the vineyards to protect against spring frost

Hail strikes later in the year and is not limited to as narrow a window of danger: June, July and August all carry its threat. In terms of physical destruction the damage caused by hail is worse and more obvious than that of frost. Shoots and leaves are broken off and litter the ground, in some cases being washed out, along with mud and stones, onto adjacent roads. It strikes suddenly and with apocalyptic force, the wind building and the sky darkening to purple-grey before the hailstones descend, usually marble-sized but sometimes the size of golf balls or bigger. A shotgun fired into the vines could hardly be more ruinous. Hail netting and cloud seeding with silver iodide are two measures used to combat hail, though both are limited in what they can achieve.

Public awareness of savage storms is greater than ever thanks to social media, though whether posting alarming photos and reports serves any useful purpose is questionable. Such reportage is one-dimensional and lacks nuance or insight; the message must be kept simple, the smartphone pointed towards the worst affected area, not turned through 180 degrees to show the untouched vines in a next door vineyard. The picture may be in colour but the message is in black and white.

OTHER CHALLENGES

For as long as I have visited the Côte d'Or I have heard, whether by way of dark muttering or voluble exasperation, that the prices being paid for vineyard land simply could not be justified. Such prices were unsustainable and there was no possibility of a return on investment, no matter how much was subsequently charged for the wine. Yet the prices continued to gallop upwards, almost every year reaching new peaks that would have been dismissed as fanciful the previous year. The temptation to sell out must be strong, which opens the possibility of an era of ping-pong trading, where small patches of vineyard are bought and sold as investments with winemaking as an afterthought.

The current losers are the small family domaines whose modest holdings may be worth millions, saddling their children with an inheritance tax bill that they may only be able to meet by selling all or part of their vineyards. It may be argued that the high prices the wines now fetch have allowed domaines to invest in new equipment and premises, which is true, but it must also be remembered that only a small number of domaines are in the pricing super league and even those whose wines end up on the auction merry-go-round do not receive anything from the subsequent resale of their wines.

The current pedestal of near-sanctification upon which the best *grands* and *premiers crus* from a select 'golden circle' of producers have been placed is impossible to explain in rational terms. Prized bottles are now treated as vinous jewels and displayed like works of art or hunting trophies, at which point they become traded commodities rather than consumable items. The term 'bottle of wine' should mean the latter, something to be anticipated, perhaps treasured for a decade in the cellar, before being enjoyed and, in that enjoyment, destroyed. Vicarious pleasure may be got from the unopened bottle but consummation is only achieved by destruction.

The prices paid at auction for back vintages from the likes of Romanée-Conti, Roumier, Rousseau, Leroy, Coche-Dury, de Vogüé

282 CÔTE D'OR

and others hog the limelight and give an invalid impression that Côte d'Or burgundy is now for millionaires only. Press releases from auction houses regularly trumpet their results, as if parting with a ludicrous sum for a comestible item is some sort of accomplishment, on a par with sporting prowess or literary greatness, rather than a form of madness.

It baffles me when high auction prices are reported in consumer publications as somehow a good thing, an achievement. Why good? Why an achievement? The readers are surely people who want to drink the wine; it's only good if you are a seller and it's likely that there are far more buyers and drinkers amongst the readership than sellers. To them, high prices are bad news, not good, and maybe now and again they should be reported as such. The genuine good news is that high quality Côte d'Or wines at reasonable, not cheap, prices can be found with a little diligent searching. The relatively new appellation, Bourgogne Côte d'Or, is a trove of vinous delights. And, in addition to its appeal on palate and wallet, it also comes with a hidden benefit that the trophy wines do not possess: it is highly unlikely to be faked.

Today's post-truth world of fake news, alternative facts and outright lies facilitates fraud. Authenticity is a debased coinage. Selling fraudulent wine in such circumstances is akin to shooting fish in a barrel. Many people don't even realize they have been conned and those who do are understandably reluctant to admit so publicly. It is easier to slip the wines back into circulation via the auction route with fingers crossed they won't be rumbled before being sold on. The boom in wine counterfeiting is also a logical consequence of ludicrous prices. If they were to plummet that would stop the counterfeiters, though it would not rid the world of all the stuff that is already in circulation.

ON A BRIGHTER NOTE

The blinkered Burgundian of yore, whose horizons stretched no further than the next village, has been consigned to the history books, though he lives on in stereotype. Today's generation of vignerons has travelled and tasted widely and, come harvest, many of them welcome *stagiaires* from around the globe to assist in vineyard and winery. They are open to innovation and development in the wine world in a way their forebears were not, the only concern being that too open a mindset may lead to increasing homogenization of styles. One of the Côte d'Or's greatest attractions is the fantastic variety and range of wines produced there. It would be a travesty if that were lost.

As well as offering more variety than expected, the côte presents an air of unchanging permanence, yet below the surface is in a state of flux. The map of vineyard ownership changes like a slowly turning kaleidoscope, with the pieces aligning and re-aligning constantly. In this century high-profile sales of domaines such as Clos des Lambrays, Bonneau du Martray, Château de Meursault and Château de Pommard have made headlines.

Worry has been expressed that such sales might open the floodgates, give 'permission' to others to sell up and initiate the most radical change in the côte's ownership structure since the Revolution. The selling off of the vineyards as *biens nationaux* at the end of the eighteenth century, allied to the fragmentation engendered by the *Code Napoléon*, which dictates that all children inherit equally, set the land ownership pattern that was to pertain for the next two centuries. But now, change is in the air again, driven mainly by prices. It is too early to say but perhaps we are witnessing the beginnings of a reversion to the pre-Revolution model when the land was owned by big 'corporations', the church and the aristocracy, but rented out to and worked by the local populace. Prestigious domaines being sold to wealthy outsiders may presage the return of an era that was once considered gone forever.

Meanwhile, the process of remorseless fragmentation continues, as inheritances are divided and redivided across the generations. With prices as high as they are, minute sub-division makes economic if not administrative sense. Such division is not a one-way street, however; pulling in the opposite direction are other forces such as when one sibling buys out others over a period of many years or when two inheritors from different domaines marry and amalgamate their inheritances to create a new domaine.

BEYOND THE CÔTE

As land prices in the Côte d'Or moved relentlessly upwards in recent decades some of the most revered names established winemaking operations beyond the narrow confines of the golden slope. The sky-high prices being paid for slivers of prime vineyard was the driving force, yet the attraction of escaping from the fish bowl of the côte to try their hand in less restrictive circumstances is also a factor. In many cases they have expanded south, to the Mâconnais, to Beaujolais and to the Languedoc,

284 CÔTE D'OR

but some have gone further afield to other continents. What follows is not meant to be a comprehensive listing, nor are their activities examined in depth, as they lie outside the scope of this book, but it is a trend worth noting as it will probably accelerate in years to come.

Beaujolais exerts a strong attraction, as indicated by Louis Jadot's purchase of Château des Jacques in Moulin-à-Vent in 1996. More recently, Frédéric Lafarge and Thibault Liger-Belair have set up there. Domaine Leflaive and Comtes Lafon are well established in the Mâconnais while, further south, a quartet of others are all making impressive wines: Domaine Dujac at Triennes in Provence, Jean-Marc Boillot in Pic Saint-Loup, Anne Gros and Jean-Paul Tollot in Minervois, and Sylvain Morey in the Lubéron. Elsewhere, Guillaume d'Angerville has expanded into the Jura, not much more than an hour's drive from his home base in Volnay. Boisset owns wineries too numerous to mention outside the côte, including California, and Joseph Drouhin is a leading producer in Oregon. Also in Oregon, Nicolas-Jay is a joint operation between Jean-Nicolas Méo of Méo-Camuzet and Jay Boberg, while in Sonoma, Faiveley owns a majority share in Williams Selyem. In Napa, Hyde de Villaine is a joint venture between the Hyde and de Villaine families who are related by marriage. Thousands of miles in the other direction Philippe Colin owns the Topiary winery in Franschhoek, South Africa, while in Gippsland, Australia, Jean-Marie Fourrier fronts a group that bought Bass Phillip in 2020. Finally, Etienne de Montille is now making wine in Hokkaido, Japan.

The micro-négoce

Côte d'Or headlines may be dominated by news of mega-deals involving tens of millions of euro, but right at the other end of the spectrum there has emerged a small cohort of tiny producers: the micro-négociants. These are niche players, buying-in grapes to produce perhaps three barrels of this wine or five of that. They barely count in the overall picture yet they add colour and interest, particularly for those Burgundy lovers obsessed with minute detail and who always want to seek out the latest, limited production treasure. In general, their wines are fresh and exciting, lively and engaging, the middle of the road is not for them. Nor are they put off by bureaucratic hurdles that could stop a runaway elephant. Look out for the following trio: Jane Eyre, Nicholas and Colleen Harbour, and Andrew and Emma Nielsen.

LOOKING AHEAD

Gaining a sense of historical perspective when writing about contemporary events is always difficult, though it helps when trying to foresee the future. To guess what the next century might bring it helps to look back an equivalent distance. One hundred years ago the depredations of the First World War were still fresh in the memory, and the United States was in the grip of Prohibition, cutting off that vital export market. Looking forward then, not even the most astute soothsayer could have foretold that only a few years later an even greater war would seize the planet. Or that a dismal half-century lay ahead for the wine trade, buffeted by outside forces such as economic depression, and from within by a less than quality-conscious approach in vineyard and cellar. Or that the century would end with an unprecedented flourish, as the wines rocketed in price and popularity.

Today, global upheaval by way of war in Ukraine and the Middle East, along with the rise of far right parties in Europe and beyond, makes trying to predict what lies ahead a trickier business than ever. It might be a futile exercise but it does have merit, for it forces us to look hard at today, thus gaining a better understanding of how things stand now. World affairs are in a state of flux and Burgundy is not immune to their influence, a circumstance amplified by geographical situation. Burgundy sits at the heart of western Europe, open to the continent, as evidenced by the huge number of visitors' cars with foreign licence plates.

For much of the period since the Second World War ended the EU has held Europe together, acting as a pacifying force between historic enemies. Any weakening of the EU bulwark, such as through Brexit, might open the door for an escalating conflict that will spiral out of control in the way Europe did in late summer 1914. If so, the Côte d'Or will not remain unaffected. It was ever thus; the story of the côte could not be told without reference to the events that have shaped it such as the Revolution, economic vicissitudes and war. The Revolution stamped the côte permanently, its grim reality now buffered by the passage of two centuries. Bloody and brutal as it was it ushered in what is probably the côte's most special epoch. Out of the upheaval came the vineyard mosaic we have today, the source of endless fascination, confusion and enchantment. If another upheaval of whatever stripe comes along in the next century it will eventually become stitched into the ongoing story of the Côte d'Or. The fabled slope will ride that wave, however grim it may be.

286 CÔTE D'OR

A chapter in the côte's future story will reference the increased competition from the likes of Germany, New Zealand, Australia, South Africa and America, north and south, with attractive prices and ever-improving wines. This will certainly have an effect and may cause consumers to drift away from Côte d'Or wines at *village* level. Whether it dims enthusiasm for the big name wines remains to be seen, so far they have been largely impervious to competition and because of small production will always enjoy scarcity value. Their allure is unlikely to fade. It is half a century since the 'Judgement of Paris' tasting shook up the wine world, and yet the French classics are still regarded as the benchmarks, still the most sought-after.

Will there be a revolution in viticulture? Perhaps. The high-yielding vines planted in the decades following the Second World War now need replanting, affording the vignerons the opportunity to rectify the errors of the past. However, they need to be mindful of the adage that 'generals always fight the last war' as they choose their rootstocks and clones and planting densities, and perhaps even plant on a north–south axis rather than the traditional east–west. They're unlikely to plant Syrah, but if climate change continues apace their descendants might rue the fact that they didn't. Thus far, climate change has been a positive force, prompting earlier, riper harvests. Whether freak weather events can be attributed to it is not certain. Frost and hail are age-old enemies; they are occurring more frequently now but their devastating effects have been experienced for centuries.

An existential threat

At time of writing alcohol is being demonized in a manner akin to that which led to Prohibition in the United States in 1920. What many do not realize is that the temperance movement had been building for decades before that; Maine was the first state to go 'dry' in 1851 and dozens were already dry by 1920. Today's anti-alcohol zealots may say that an outright ban is not their aim but they cannot speak for their successors who, in 30 or 40 years' time, may give thanks for the groundwork done by today's temperance brigade, allowing them to push for draconian restrictions on alcohol. Prohibition validated law-breaking by otherwise law-abiding citizens and the USA has never fully recovered from the mayhem it begat. We are not yet close to that apocalypse but the recent declaration by WHO that there is no safe level of alcohol consumption may be a straw in the wind.

The vignerons' deep-seated connection to the land built up over generations may be sundered as more estates fall into foreign or large-corporation ownership. A manager appointed for managerial skills will not have the same innate knowledge as the local farmer and perhaps the winemaking philosophy will be dictated from the accountant's desk and not the *enjambeur* of the vigneron. The changing structure may be a good or a bad thing; at first the natural instinct is to decry such change but the pixelation of vineyard ownership and the convoluted web of domaine ownership today is not without its problems. In many cases 'owners' of a family domaine will have siblings in the background from whom they rent their portion of vineyards, or whom they are striving to buy out. Move on a generation and there could be a dozen cousins, not all of whom have a strong connection to the domaine. Move on another generation and there might be dozens of cousins, perhaps scattered across the globe, with no interest in the domaine except for the annual dividend. If a domaine is thus hamstrung it cuts off investment and can lead to stagnation, so selling out may be beneficial for all concerned, and for wine quality.

Wine quality in the Côte d'Or has been on an upward curve for four decades thanks to improvements in vineyard and winery, where one piece of equipment is singled out for credit: the sorting table. It is now *de rigueur* for a quality-conscious domaine to sort their grapes at least once and many do it twice, sometimes in addition to a preliminary sorting in the vineyard. Such rigour sees near-perfect grapes going into the vats but it is tempting to wonder how far super sorting can go before it becomes counterproductive. Might the wines end up as photoshopped caricatures of their true selves? Will the pursuit of polished perfection unravel in the face of a demand for character ahead of flawlessness? Suggesting there might be a retreat from super sorting in the coming decades may sound like heresy but it is salutary to remember that it is often the most enshrined practice or belief of today that is modified tomorrow.

CONCLUSION – A GOLDEN AGE?

In today's Côte d'Or there is a legion of talented and conscientious winemakers bringing extraordinary levels of diligence, perhaps never seen before, to getting the most out of their vineyards. Visiting scores of them to research this book, though enervating at times, was a constant

288 CÔTE D'OR

source of stimulation and validation of my belief in their resilience. In years to come historians may say this was a golden age when artisanal wines, almost crafted individually, reached their peak. These winemakers are the bespoke shirt makers and shoemakers of the wine trade, existing in stark contrast to all that is produced on an industrial scale. Their future survival may be determined by outside forces, just as the Revolution and phylloxera forged the côte of today. Perhaps it is now sliding into the grip of two equally strong forces: spiralling land prices and climate change. It is impossible to say how they will affect the Côte d'Or, but in the long term it will be enriched by whatever befalls it. Future generations will be as enchanted by it and its wines as all others have been for centuries.

Côte d'Or. Slope of Gold. A two-thousand-year backstory of human endeavour is captured in the name of this vineyard sliver. Over the centuries all manner of depredations have been thrown at it, and blessings showered on it, and through it all the côte has endured. There's magic in those hills; that's immutable, though harnessing it hasn't always been easy. War and pestilence, depression and revolution, all have played their part in hampering the growing of grapes and the making of wine, yet wine has always been made. Whatever ups and downs the Côte d'Or has experienced over the centuries, it has endured.

8

VISITING THE CÔTE D'OR

Only when a name such as Gevrey-Chambertin or Chassagne-Montrachet is seen on a road sign at the entrance to a village does the Côte d'Or truly come alive. The jumble of wine names previously known solely from labels and wine lists resolves into a sequence of real places with fixed geographical identity. Seeing the Hill of Corton for the first time or strolling the hillsides above Vosne-Romanée or Puligny-Montrachet, one cannot help feeling perplexed that a strip of agricultural land, albeit one that is tended and manicured like no other, can be the source of such vinous fame and fable. Visit for a week or two and that feeling of perplexity will give way to one of enduring fascination. Placing your feet on the ground is the way to gain a true understanding of the Côte d'Or.

The côte is open to visitors like never before. Visiting today is a more pleasant, more rewarding and more easily arranged exercise than heretofore. That said, it is not your average holiday destination so some care and thought must be put into planning a visit if it is to prove rewarding. For obvious reasons summer is probably the best time to go but a visit to the côte at any time of year always brings rewards. If you don't mind the cold, January's Saint-Vincent Tournante festival is a must-do.

SAINT-VINCENT TOURNANTE

www.tastevin-bourgogne.com

Saint Vincent is the patron saint of wine and winemakers and effigies of him, some simple, some elaborate, can be found in many Côte d'Or cellars, usually tucked into a niche from where he looks over the maturing barrels. The annual festival in his name takes place on the final

The défilé (parade) at the start of the Saint Vincent Tournante, Puligny-Montrachet 2022

weekend of January and, *tournante* meaning revolving, it is held in a different village every year. It was founded by the *Confrérie des Chevaliers du Tastevin* in 1938, four years after that brotherhood's foundation, the aim of both being to raise the profile of Burgundy and celebrate all that is good in the wine. Both have succeeded marvellously well.

The *confrérie* now has thousands of members worldwide and the *tournante* attracts tens of thousands of visitors and participants every year. The host village goes to extraordinary lengths to deck itself with multicoloured paper flowers and every manner of decoration and display, turning the grey of winter into technicolour spring. A parade through the vineyards starts in darkness with an early morning glass of wine and ends with Mass in the church. Thereafter, any number of food and wine outlets, along with numerous other entertainments, help to keep the cold at bay. The *tournante* is a weekend-long, village-wide street party that captures the spirit of Burgundy at its finest.

LES GRANDS JOURS DE BOURGOGNE

www.grands-jours-bourgogne.com

For anyone in the wine trade keen to brush up on Burgundy by way of full immersion therapy this is the event to attend – but beware, this is wine tasting boot camp, a week-long marathon that taxes the hardiest of palates. Held every second year in March, *Les Grands Jours* is a superbly well-organized shop window for Burgundy; hundreds of wines can be tasted and dozens of producers met, in a series of different locations. As a Burgundy snapshot it is hard to beat.

Rooms with a view

Time was when a cellar tasting in the Côte d'Or was conducted in subterranean, cobwebbed gloom, with tall heads stooped beneath the low, arched stonework to avoid injury. Increased prosperity in recent years, however, has seen a host of spanking new tasting rooms constructed to receive visitors, some fitted out like luxury lounges. Three spring to mind, the common thread linking them being an evocative view over the vines whence your wine came, as you sip your sample. At Clos de Tart in Morey-Saint-Denis the gable-end window looks directly onto the *monopole clos*, the only *monopole grand cru* outside Vosne-Romanée. Meanwhile, in Chassagne-Montrachet, Bruno Colin's slick new tasting room overlooks the *premier cru* La Maltroie, so close you feel that you could reach out and touch the vines. And in Puligny-Montrachet, at Etienne Sauzet, another gable-end window, broad and arched, faces the setting sun as it dips below the bare hilltop, or *mont rachet*, from which Montrachet takes its name.

The new tasting room at Domaine du Clos de Tart

292 CÔTE D'OR

BEAUNE

www.beaune-tourism.com

People either like Beaune or they do not, there is no middle ground. Some dismiss it as a tourist trap, but it traps plenty of locals too. It is a wonderful place for a wander and if the bustle in the Place Carnot and surrounding streets gets too much then walk for a minute in any direction to discover curving, near-silent back streets with discreet brass plaques of *notaires* and suchlike. Unless you have a compelling reason to stay elsewhere, Beaune is the best place to base yourself for a visit to the Côte d'Or. Stay as centrally as possible – the Hotel Le Cep is a gem for wine lovers, with a tasting cellar opened in 2017, now augmented by a cigar lounge. When the day's exploration is finished a score of restaurants will be within easy walking distance, saving some poor soul from designated driver penance, never mind the fiendishly expensive taxis. And do not miss the Athenaeum book, wine and wine accessories shop. Standing opposite the entrance to the Hôtel-Dieu, it stocks a superb selection of wine books, glassware, maps, souvenirs and a reasonable number of wines. Hours can be whiled away browsing there.

Hôtel-Dieu

The Hôtel-Dieu was built by Nicolas Rolin, chancellor to Duke Philip the Good, in 1443 to house the Hospices de Beaune, a charitable institution that provided not just medical care but also support for the indigent citizenry who were suffering from the ravages of the Hundred Years' War and the plague. The institution itself moved to modern premises on the outskirts of Beaune in 1971, leaving the splendid fifteenth-century buildings, ornately roofed in polychrome tiles, to stand as the Côte d'Or's single most impressive edifice. The exterior façade is relatively plain, in contrast to the ornate courtyard within, where elaborate half-timbered gables vie for your attention with the roofs' dazzling tile patterns. The most impressive interior spaces are the *salle des pôvres* (room of the poor) and the chapel, which abut one another so that the patients, accommodated two to a bed, could easily hear Mass. There is a strong Flemish influence in the building, underscored by Rogier van der Weyden's spellbinding *Last Judgement* that once sat above the altar and is now housed in the hôtel's museum.

Hospices de Beaune wine auction

www.hospices-de-beaune.com

The most famous charity wine auction in the world was founded in 1859 and raises money for the hospital that was once housed in the Hôtel-Dieu. Since 1459 the Hospices has been the recipient of vineyard donations, the produce of which it has traditionally sold off to fund its activities. Today it owns some 60 hectares of vineyard and the auction is held on the third Sunday in November every year. In 2024, some 439 barrels of wine sold for nearly €13.9 million (US$14.25 million). Depending on whether you are involved or not, it is as exciting or boring as any auction, though the scrum that precedes the off, as the dignitaries make their way to the podium, is worth seeing. Cameramen jostle for position, the intensity of their efforts depending on the fame of the year's honorary presidents.

The auction takes place in the covered market hall, which is splendidly decked for the occasion and is completely unrecognizable as the functional space occupied by butchers, fishmongers and cheesemongers on market day. Over the course of the weekend Beaune feels invaded by the world's wine trade and interested tourists; there's a slightly manic air, a rush hither and thither. Endless speculation, tale telling and anecdote swapping generate a lot of hot air. It is a wonderful event and all in a good cause, but I prefer Beaune at almost any other time of year. The auction forms part of *Les Trois Glorieuses*, a weekend of celebration that also includes a grand dinner at Clos de Vougeot and an epic lunch, *La Paulée de Meursault*, for those still strong in wind and limb.

Cité des Climats

www.citeclimatsvins-bourgogne.com

Opened in 2023, the Cité on the outskirts of Beaune is one of three new visitor centres in Burgundy, the others being in Chablis and Mâcon. Writing at the time of opening I described it thus: 'The *Cité* is a magnificent construction, given an almost weightless appearance by the sweeping external ramp that spirals upwards and which can be accessed without having to pay an entry fee. Inside, there is an impressive exhibition space, a wine school and a top floor wine bar with views back towards Beaune and the vineyards beyond.' Its location on the outskirts of Beaune means that it has to compete with town-centre attractions to woo visitors but for newcomers to the region it is worth the effort. Others may wish to concentrate their efforts on vineyard visits.

The new Cité des Climats et vins de Bourgogne in Beaune

WORLD HERITAGE

The Côte d'Or celebrated in July 2015 when, after an application process that lasted nine years, it was inscribed on UNESCO's World Heritage list of sites of cultural and historical significance. The argument ran that the côte is a unique example of a vineyard region where the potential to make great wine existed, but that that potential was and is constrained by vagaries of climate and soil and it has only been harnessed thanks to mankind's intervention. Natural conditions harnessed by human will are what make it special.

Since its elevation the côte has received a face lift, most visibly in the restoration of the traditional dry stone walls that mark the vineyard boundaries and which, when running across the slope allow surface water to pass through while reducing soil erosion. There is now the money and the will to do such work where, for over a century after phylloxera, the côte was marked by decline. A century's neglect will take time to eradicate and it is good to see traditional practices and methods being reinstated, but it is also important that there is no attempt to set the côte in aspic, that it is not turned into a living museum or a theme park thanks to the UNESCO status.

Hand in hand with UNESCO status was the opening of the Maison des Climats in July 2017. Housed beside the tourist office at the Porte Marie de Bourgogne in Beaune, this makes an ideal starting point for an exploration of the Côte d'Or, especially for first-time visitors. It offers a snapshot immersion in the history, geology and culture of the côte, and a fifteen-minute visit suffices to see the audio-visual presentation, enlivened by superb photography.

LA MAISON VOUGEOT

www.lamaisonvougeot.com

Hollywood meets Burgundy in a way never seen before in Vougeot's La Maison, opened in 2016. It is the creation of Jean-Charles Boisset and his sister Nathalie and it is very much in his flamboyant image. A succession of spaces runs from cosy to challenging, the whole place dazzles and intrigues, lights flash and sparkle, sound effects gush and whisper. Wines can be sampled and jewellery bought. There's a lush theatrical feel here, everything is svelte and smooth. For a totally new and non-traditional visitor experience in the Côte d'Or this is the place to see.

VOSNE-ROMANÉE

Vosne-Romanée may appear on first acquaintance to be the most buttoned-up of all the Côte d'Or villages but it has become noticeably more visitor friendly since the first edition of this book. 'I wanted people to be able to drink a glass of Vosne in Vosne. Up to the time we opened, it was easier to drink a glass of, for instance, Echézeaux in Hong Kong than in Vosne-Romanée where it is made.' Louis-Michel Liger-Belair is speaking of his new wine bar, La Cuverie (lacuveriedevosne.fr), described on the sign outside as, 'Épicerie – Café – Bar à Vins – La Poste'. Yes, it includes a diminutive post office but the appeal here is in the huge range of wines; every Vosne domaine is welcome to supply their wines for sale here.

Next-door neighbours, Domaine Mugneret-Gibourg, have converted the original family home into a five-bedroom boutique hotel, La Maison de Jacqueline (www.lamaisondejacqueline.fr), approached along a driveway that runs through La Colombière, a *village* vineyard of which 20 rows had to be sacrificed to create space for a pool, garden

296 CÔTE D'OR

and parking. It is named after the late matriarch of the domaine and managed by her granddaughter, Fanny.

Both opened in 2022, both are worth visiting.

THE VINEYARDS

It strikes me that far too many visitors to the Côte d'Or spend the greater part of their time in producers' cellars, with precious little spent exploring the vineyards. There is nothing to beat a walk amongst the vines, ending at the tree line above the *climats* and *lieux-dits* and looking back across the slope. A favoured spot for this exercise is the track that separates Ruchottes-Chambertin from the forest. Looking directly down the slope, the eye travels across Ruchottes to Mazis-Chambertin then sweeps right to Clos de Bèze and left to the village of Gevrey-Chambertin. Or stand at the cross placed at the apex of the hill of Corton and take in the great swathe of vineyard from Pernand-Vergelesses, away to your right, across Savigny-lès-Beaune and around to the village of Aloxe-Corton.

A night in the vines

Sleeping rough in a vineyard may be some people's idea of fun, communing with the vines while wrapped in the arms of Morpheus, but for less hardy mortals the idea has little attraction. Step forward La Folie de Vougeot (www.la-folie-vougeot.com), a luxury, one-bedroom retreat in the heart of Clos de Vougeot, the expansive *grand cru* that can rightly claim to be the historical heart of the Côte d'Or. It may also be booked for tastings and meals, where you can bring your own wines, or choose from those of the co-owners, Maison Joseph Drouhin and Domaine de Montille.

Even if time is short, resist the temptation to rush hither and thither, burning up the pixels as you go, capturing it all on camera and video but seeing little and taking less in. Stop and stand, marvel and muse. If you have travelled from the other side of the world on a once in a lifetime visit that is hard advice to follow but ponder this: going at breakneck speed will yield little more lasting knowledge of the vineyards than watching a few films about the region and checking some websites. Where the exposure of a slope turns or an incline changes can only be gleaned by walking that slope.

I always counsel a preference for vineyard exploration over cellar visits. Too often, schedules are structured in reverse of this, yet, atmospheric as an ancient cellar can be, the wine you taste there can probably be tasted back at home. The wine travels, the vineyard doesn't. To be able to muse, months or years later, while drawing the cork on a cherished bottle: 'I stood on the ground whence this wine came' carries a deeper ring than 'I met the vigneron who made it'. Best of all is to explore a vineyard with the vigneron while sipping a glass of wine from that vineyard.

THE MAPS

As an aid to exploring and understanding the Côte d'Or the maps of Sylvain Pitiot and Pierre Poupon are unparalleled. They reward close examination, to follow a contour line as it weaves across the land, favouring one vineyard with perfect exposure, hobbling another with late afternoon shade, or to winkle out the name of a remote *lieu-dit*. But they should be stood back from also, when they yield an insightful overview and reveal the regimented neatness of the Côte de Nuits when compared with the Côte de Beaune's ragged sprawl. The Côte de Nuits, especially south of Gevrey-Chambertin, is a neat progression of communes, tightly focused with little deviation from the linear march southwards. The Côte de Beaune is more haphazard, spilling across the map this way and that; even the extra colours add flamboyance, a Hawaiian shirt to the northern cousin's pinstripe. The maps inadvertently capture something of the contrasting character of each côte, the northerners more reserved, the southerners more gregarious. No attempt to learn about the Côte d'Or should be made without constant reference to them. No book, no film, no blog, no monograph, no essay, no doctoral thesis, no erudite treatise, nothing can reward repeated examination and perusal as they do. There is always something new to be learned, some conundrum to be resolved, and another to be spotted for the first time. This brilliant resource is also available as a smartphone app, ClimaVinea, which is the perfect aid to vineyard exploration, as it tells you precisely where you are on the map, so you'll never get lost.

GLOSSARY

Are. A hundred square metres.

Argilo-calcaire. Clay-limestone, the classic foundation of the Côte d'Or vineyards.

Barrique. Barrel, the standard Burgundy barrel is 228 litres. Known as a *pièce* in the Côte d'Or. Larger barrel sizes are becoming increasingly common.

Bâtonnage. Lees stirring in barrel, most usually done with white wines. Less popular now than in the past.

Biens nationaux. National goods or assets. Specifically refers to property seized from the church and aristocracy during the French Revolution and then sold.

BIVB. Bureau Interprofessionnel des Vins de Bourgogne. The producers' association, tasked with research and generic promotion.

Climat. A named vineyard plot. Perhaps the most confusing term associated with the Côte d'Or. It does not translate directly into 'climate' in English. See box, page 24, *Climat* or *lieu-dit*?

Clos. A vineyard historically enclosed by walls on at least three sides. Many still are walled but others where the walls no longer exist are still designated *clos*.

Combe. A side valley that usually runs roughly perpendicular to the main line of the Côte d'Or.

Côte. Slope or hillside, not the whole hill.

Courtier. A middleman who brokers deals between those with grapes to sell and those who want to buy them. The buyers were traditionally the large négociants but are now also the micro-négoces. Increasingly important in times of short supply.

Cru. Usually translated as 'growth'. Not quite equivalent to *climat* or *lieu-dit* but is used to refer to a vineyard's appellation controlée status, e.g. *grand cru*, *premier cru*.

Cuvée. A quantity of wine, a vat or a blend of several vats. Its name may match that of a *cru* or it may be named by the vigneron if, for instance, it is a blend of several *crus*.

Délestage. The removal of all the liquid from a fermenting vat before splashing it back onto the cap of skins. Also called rack and return.

Demi-muids. Traditionally a large barrel of 600 litres. Today refers to barrels of 350, 400 and 500 litres increasingly used for white wine.

Domaine. An estate, often family-owned. A domaine wine is made from vines owned by the estate. The distinction between domaine and négociant is not as clear-cut as previously.

Élevage. The rearing or raising of wine in the cellar after fermentation, usually in barrel followed by some time in tank.

En foule. In a crowd, the way vines were planted before phylloxera, usually at a much higher density than today.

Lees. The sediment deposited at the bottom of a barrel or tank after fermentation.

Lieu-dit. A named place or vineyard plot. Has almost the same meaning as *climat*. See box, page 24, *Climat* or *lieu-dit*?

Lutte raisonnée. Might best be translated as 'reasoned response', though 'struggle' would be more literal. A slightly woolly term used to describe the methods and philosophy employed by non-organic vignerons who apply chemicals as needed and not as a matter of routine.

Marc. The solid cake of pips, skins and stems left in a vat after the wine has been run off, subsequently distilled into the spirit *Marc de Bourgogne*.

Millerandage. Uneven fruit set yielding berries of different sizes. Can give beneficial concentration in the resultant wine.

Monopole. A vineyard with only one owner. Rare, and usually prized.

Murgers. Also *meurgers*. Piles of rock and stone removed from the ground over centuries by the vignerons, to prepare it for planting, and found at the perimeter of vineyards.

Must. The grape juice prior to fermentation.

Négociant. Traditionally a merchant who bought grapes, juice or wine, used to produce wines under their own name. Many now also have substantial vineyard holdings.

300 CÔTE D'OR

Oidium. A fungal disease, also known as powdery mildew, that inhibits photosynthesis.

Ouvrée. A traditional measurement of vineyard area, the amount that could be worked by a man in one day. There are about 24 *ouvrées* in a hectare. Still used in land transactions.

Phylloxera. The louse that devastated European vineyards in the second half of the nineteenth century by eating the vines' roots. Combated by grafting onto resistant American rootstocks.

Pigeage. Punching down. In red wine fermentation, the plunging of the cap of skins into the liquid below to extract colour and flavour.

Provignage. The system of propagating vines by layering prior to phylloxera.

Racking. Moving wine from one barrel to another to aerate it and remove lees.

Remontage. Pumping over. Drawing the liquid from the bottom of a fermenting vat and pumping it onto the cap of skins to extract colour and flavour.

Repiquage. Replanting vines individually, rather than in blocks.

Table de tri. Sorting table. Today, an indispensable piece of equipment that acts as the gatekeeper for grapes between vineyard and winery.

Tastevin. A broad and shallow metal tasting cup with dimples and grooves to reflect the light and show the colour and clarity of a wine. Seldom used today.

Tonnelier. Barrel maker or cooper.

Tracteur enjambeur. Tractors that straddle the vine rows, with a wheel on each side. They look ungainly anywhere except in a vineyard. The giraffes of the tractor world.

Vendange. Harvest.

Vigneron. Best translated as 'wine grower'. A winemaker. Often used interchangeably with *viticulteur*.

Viticulteur. Traditionally, *viticulteurs* were vineyard owners whose primary role was vineyard work.

BIBLIOGRAPHY

Ayscough, Aaron, *The World of Natural Wine: What It Is, Who Makes It, and Why It Matters*, Artisan, New York 2022.

Blake, Raymond, *Breakfast in Burgundy: A Hungry Irishman in the Belly of France*, Skyhorse, New York 2014.

Blake, Raymond, *Wine Talk: An Enthusiast's Take on the People, the Places, the Grapes, and the Styles*, Skyhorse, New York 2022.

Brook, Stephen (General Editor), *A Century of Wine: The Story of a Wine Revolution*, Mitchell Beazley, London 2000.

Cannan, Christopher, *A Journey in the World of Wine: From Operating in the Trade to Founding a Boutique Winery*, independently published 2023.

Clarke, Oz, *The History of Wine in 100 Bottles: From Bacchus to Bordeaux and Beyond*, Pavilion, London 2015.

Clarke, Oz, *Oz Clarke on Wine: Your Global Wine Companion*, Academie du Vin Library, London 2021.

Coates, Clive, *Côte d'Or: A Celebration of the Great Wines of Burgundy*, Weidenfeld & Nicolson, London 1997.

Coates, Clive, *The Wines of Burgundy*, University of California Press, Los Angeles 2008.

Gambal, Alex, *Climbing the Vines in Burgundy: How an American Came to own a Legendary Vineyard in France*, Hamilton Books (Imprint of The Rowman & Littlefield Publishing Group), New York 2023.

Garcia, Jean-Pierre (trans: Colas, Maxine), *The 'Climats' of Burgundy: A Unique Millennia-Old Heritage*, Glénat, Grenoble 2013.

302 CÔTE D'OR

Gotti, Laurent (trans: Bobes, Stevie), *The Grand Crus of Burgundy: Detailed Atlas and Characterization of the Climats*, Collection Pierre Poupon, Beaune 2023.

Goulding, Henry, *Thoughts on a Wine Cellar: A Love Song*, independently published 2022.

Hanson, Anthony, *Burgundy* (third edition), Mitchell Beazley, London 2003.

Healy, Maurice, *Stay Me with Flagons*, Michael Joseph, London 1941.

Howkins, Ben, *Adventures in the Wine Trade: Diary of a Vintners Scholar*, Académie du Vin Library, London 2023.

Jefford, Andrew, *The New France*, Mitchell Beazley, London 2002.

Jefford, Andrew, *Drinking with the Valkyries: Writings on Wine*, Academie du Vin Library, London 2022.

Johnson, Hugh, *The Story of Wine*, Mitchell Beazley, London 1989.

Johnson, Hugh, *Wine: A Life Uncorked*, Weidenfeld & Nicolson, London 2005.

Keevil, Susan (Editor), *On Burgundy: From Maddening to Marvellous in 59 Wine Tales*, Academie du Vin Library, London 2023.

Kladstrup, Don & Petie, *Wine & War: The French, The Nazis, and France's Greatest Treasure*, Coronet, London 2001.

Landrieu-Lussigny, Marie Hélène & Pitiot, Sylvain (trans: Dent, Delia & Renevret, Françoise), *The Climats and Lieux-Dits of the Great Vineyards of Burgundy: An Atlas and History of Place Names*, Meurger, Vignoles 2014.

Loo, Bart van (trans: Nancy Forest-Flier), *The Burgundians: A Vanished Empire*, Bloomsbury, London 2021.

Morris, Jasper, *Inside Burgundy: the Vineyards, the Wine and the People*, Berry Bros & Rudd Press, London 2010.

Morris, Jasper, *Inside Burgundy* (second edition), Berry Bros & Rudd Press, London 2021.

Morrison, Fiona, *10 Great Wine Families: A Tour Through Europe*, Academie du Vin Library, London 2019.

Nanson, Bill, *The Finest Wines of Burgundy: A Guide to the Best Producers of the Côte d'Or and Their Wines*, Fine Wine Editions, Aurum Press, London 2012.

Norman, Remington, *The Great Domaines of Burgundy* (second edition), Kyle Cathie, London 1996.

Norman, Remington, *Grand Cru: The Great Wines of Burgundy through the Perspective of its Finest Vineyards*, Kyle Cathie, London 2010.

Parker, Robert, *Burgundy: A Comprehensive Guide to the Producers, Appellations and Wines*, Dorling Kindersley, London 1990.

Pitiot, Sylvain & Servant, Jean-Charles, *The Wines of Burgundy* (thirteenth edition), Collection Pierre Poupon, Beaune 2016.

Rigaux, Jacky, *Burgundy Vintages 1846–2006*, Terre en Vues, Clémencey 2006.

Roux, Julie (trans: Jachowicz-Davoust, Barbara), *The Cistercians*, MSM, Vic-en-Bigorre 1998–2005.

Scruton, Roger, *I Drink Therefore I am: A Philosopher's Guide to Wine*, Continuum, London 2009.

Spurrier, Steven, *A Life in Wine*, Academie du Vin Library, London 2020.

Wilson, James E., *Terroir: The Role of Geology, Climate, and Culture in the Making of French Wines*, Mitchell Beazley, London 1998.

INDEX

Index notes: Prepositions de, d', des, and du, also articles le, la, les, are ignored in alphabetical listing. Page references printed in bold refer to boxes, and those printed in italic refer to maps and photographs.

acidity 17, 40, 61, 62
ageing
  in bottle 82, 99, 162, 176, 188, 201, 203,
    260, 269
  Gamay grape and 53
  processes 59, 93, 149, 203
Aligoté grape 52–3
  in production 67, 70, 151, 163, 182, 185,
    200, 238, 255
Aloxe-Corton 29, 37
  topography 157–8, *158,* 160
  wines 146, 252
  see also Hill of Corton
Ambeis, Franck 115
Amiot, Fabrice 228
Les Amoureuses
  topography 25, *113*
  vineyard holdings and replanting 108, 112,
    116
  wines 96, 105, 189
  premier cru 116
amphorae 189
d'Angerville, Guillaume 185–6, 284
d'Angerville, Marquis Jacques 186
Appellations d'Origine Contrôlée 15
Archambaud, Jean-Philippe 251
Arlaud, Cyprien 91–3
Armand family 180–82
Artémis group 169
Au-Dessus des Malconsorts *122,* 178
Audoin, Cyril 67–8
Les Aussy 201

Auxey-Duresses
  topography 32, 157, *158*
  vineyard holdings 195, 210
  wines 190, 192–3, 199, 211
  premier cru 180–81
Avenel, Vincent 170–71

Bachelet family 243
Barnier, Frédéric 173–5
Les Barreaux 124–5
barrels 59, **61,** 104, 109–10, 134–5, 137, 155,
    200, 228, 232, 241
Bâtard-Montrachet 239–40
  topography and terroir 34
  vineyard holdings 161, 221, 234, 244
  wines 142, 220, 222, 234
  grand cru 223
bâtonnage 62, 145, 155, 228, 253
  possible adverse effects of 278
Les Baudes 88, 101, 112, *113*
Baum, Michael 182–3
Beaujolais 284
Beaune
  cellars 169, 172
  history 5, 10, 13
  négociants 168
  producers 168–79
  topography 167–8
  visiting 292–4
  wine auction 293
  wines 168, 233
  premiers crus 171–2, 173, 176–7, 233

306 CÔTE D'OR

Beaurepaire 36, 240
Les Beaux Mont Bas 124
Les Belles Filles 241
Bienvenues-Bâtard-Montrachet 34–5, 220
biodynamic vineyards 42–4, 90, 164, 178, 224, 243
Bitouzet, François 200
Blagny 211
Blake, Raymond 2–3
Boillot, Jean-Marc 284
Boisset 144–5, 154, 202, 284
Boisset, Jean-Charles and Nathalie 145, 154, 295
Bonaparte, Napoleon 11–12
Bonet, Elodie 103–4
Bonnes Mares
    renown 24
    topography 24, 105
    vineyard and holdings 23, 69, 93, 108, 189, 213, 250
    wine 96, 100
    grand cru 69–70, 109, 110
bottling closures 62–3, 220, 223, 278–9
Les Bouchères 209
Bouchey, Patricia x
Boudet, Emilie 224
La Boudriotte 35, 226, 241
bougies (smudge pots) 279, 280
Bourgogne
    Aligoté 163
    Blanc 149, 214, 216, 223, 236, 237
    Côte d'Or Cuvee Heritage 207
    Côte d'Or 104, 259, 282
    name 21
    reputation 282
    Rouge 119, 131, 151, 202
    tri 'tripartite' wine 210
Bouygues family 84
Les Bressandes 30, 159, 165
Brochon 21, 29, 75, 78
Les Brunettes et Planchots 252
Buisson, Franck and Frédérick 191–2
Bureau Interprofessionnel des Vins de Bourgogne (BIVB) 13, 176, ix
Burgundy wine
    and 1996 premox 62–3, 222, 276–9
    ageing 273–4
    enjoying 270–71
    and food 272–3
    and glassware 272
    prices at auction 281–2
    serving 271
    shifting trade structure 275–6
    see also Côte d'Or
Burgundy/Bourgogne 21

Caillerets, Chassagne-Montrachet 35, 226
    premiers crus 228, 234, 238, 240, 241–2, 245
Le Cailleret, Puligny 35, 217, 222
Caillerets, Volnay 32, 164, 184, 187, 189
Caradeux 117–18
La Cardeuse 247
Carillon, François and family 217
Carillon, Jacques 219–20
cellars of Beaune 169, 172
Les Chaffots 118
Les Chalumeaux 210
Chambertin-Clos de Bèze 21–2, 73, 74, 81, 85, 96–7
Chambolle-Musigny
    producers 105–112, 118
    topography 37, 65, 105
    vineyards 24, 113
    wines 93, 99, 136, 137, 189
        grands crus 69–7, 189
        premiers crus 88, 96, 101, 104, 116, 134, 250–51
Champ Canet 217, 224
Champans 186–7
La Chatèniere 243–4
Chapelle-Chambertin 87
Chapier, Jean-Marie and Isabelle 76
chaptalization/adding sugar 60–61
Chardonnay
    plantings 25, 29, 32, 148, 214
    in production 161, 206
    pruning and harvesting 51–2
    renown 48
    wines
        of Meursault 196, 206, 210, 214
        of Puligny-Montrachet 218, 219
        reputation 51
Charlemagne 29, 162
    see also Hill of Corton
Charlopin, Philippe 78–9
Les Charmes, Chambolle-Musigny 69, 113
Les Charmes, Meursault 33, 217
    premier cru 206, 208, 210, 213
Charmes-Chambertin 93, 151–2
Les Charmes-Dessus 201, 213
Le Charmois 244
Chartron, Jean-Michel and Anne-Laure 221
Chassagne-Montrachet 33, 34
    producers 228–42
    renown 159, 226–7
    topography 39, 158, 158, 225–6
    vineyards and holdings 35, 154, 195, 207, 243, 248
    map of 226–7
    visiting 291

INDEX 307

wines
premiers crus 203, 214, 229, 231, 235, 237–8, 239, 240, 245, 249
rouge **50**, 258
see also Caillerets, Chassagne-Montrachet
Château de la Tour 120–21
Château de Meursault 211–13
Château de Pommard 182–4
Château de Puligny-Montrachet 213–14
Château du Clos de Vougeot 9, *34*, 37, 114, 293
Les Chaumées 35, *227*, 231
Les Chaumes 127
Les Chenevottes 35, 171, *227*, 237–8
Chevalier-Montrachet 33
vineyards 223, *257*
wines 52, 175, 189, 231
grand cru 222
Les Chenevery 101
Chevrot, Pablo and Vincent 254–5
Chorey-lès-Beaune 30, 163, 165–6, 165–7
premier cru 167
Cité des Climats, Beaune 293
Clair, Arthur 69–70
Clair, Bruno 23, 68–9
Clavoillon 223
climat/lieu-dit 24
climate and weather 45–6, 174, 279–80, 286
climate change 221, 234, 237
frost and hail *45*, 256–7
effect on vintages *see* vintages
Clos Blanc de Vougeot 155–6
Clos de la Chapelle 154
Clos de Bèze 6, 11, 21–2, 68–9, 85, 96–7
Clos de Chênes 32, *184*, 200, 209
Clos de la Caves des Ducs 178
Clos de la Chantenière 247
Clos de la Maréchale 108–9
Clos de la Perrière 25, *34, 113*, 151, 154
premier cru 115–16
Clos de la Roche
lieux-dits **24**
map *92*
vineyards and holdings 23, 93, 101, 118, 189
wines 146, 163
grand cru **19**, 95
Clos de Tart 8, 23, 91, *92*, 102–3, 291
Clos de Tavannes 36, 190, 231–2
Clos de Thorey 155
Clos de Vougeot 26, 119, 121, 126, 127–8, 131, 136, 151, 155–6, 175
classification 159
history 8, 9, 11–12, 129, 132–3
lieux-dits 26, 30
producers 120–21

renown 25–6, 64
topography 26–7, 37–8, *65, 113*, 114
vineyards and holdings 85, 103, 114, 155, 156
visiting 296–7
wines 30, 114
Clos de Vougeot Musigni **33**, 126
Clos des Chevaliers 222
Clos des Ducs 32, *184*, 185–6, 194–5
Clos des Epeneaux 180–82
Clos des Fèves 171–2
Clos des Lambrays 15, 23, 81, *92*, 97–8, 283
Clos des Mouches 31, 173, *197*
Clos des Poutures 236
Clos des Ursules 30, 174
Clos du Château des Ducs 187
Clos du Ducs 185–6
Clos du Roi 30, 159
Clos Marey Monge 182–4
Clos Pitois 35, *226*
Clos Rognet 132
Clos Rousseau **33**, 36, 249, 255
Clos Saint-Denis 93, 100
Clos Saint-Jacques 22, 69, 82–3
Clos Solon 23, *92*, 101
Les Clousots 204
Colin, Bruno 230
Colin, Damien and Caroline 244–5
Colin, Joseph 244
Colin, Philippe 284
Colin, Pierre-Yves 244
La Combe d'Orveau 25
Les Combettes 199, 219, 223, 225
'commune' definition 3–4
communes, boundary regulation 15
Confrérie des Chevaliers du Tastevin 79, 114, 146
Saint-Vincent Tournante 289–9
coopers 232
Les Corbeaux 90
corks 62–3, 278–9
Corton 26
history 141
map *158*
vineyards and holdings 115, 162
wines 29, 148, 157, 159
grand cru 131, 132, 141, 148
red 160, 162
see also Hill of Corton
Le Corton 30
Corton-Bressandes 165
Corton-Charlemagne 7, 29, 53, 142, 161–2
vineyards and holdings 115
wines 30, 157, 204–5, 214, 221, 241
see also Hill of Corton

308 CÔTE D'OR

Côte de Beaune 22, 153
  appellation **31**
  principal vineyards 29–37
    map of *158*
    producers 100, 141, 159–255
    topography 28, 142, 157–9, 297
Côte de Beaune-Villages 31, 37
Côte de Nuits 169, 189, 251–2
  geology and topography 40, 297
  principal vineyards 20–29
    map of *65*
villages and producers 64–156
Côte de Nuits-Villages 29, *65,* 250
Côte d'Or
  climate 45–6
  geology 39–41
  harvest 53–6
  history 5–17, 54, 102, 138–9, 211
    and future 285–6
  land ownership
    headline sales 283
    legislative hamstringing 287
    and migration 283–4
    purchase prices 281–2
  the Pitiot and Poupon maps 297
  renown 1–3, 281–2, 287–8
    as World Heritage site 294–5
  terroir 40–41
  topography 4, 37–9, *65*
  visiting **291**, 296–7
  viticulture 41–4
  see also Burgundy wine; Côte de Beaune;
    Côte de Nuits
courtiers 232, 276
Covid vintage 260
Les Cras 207
Criots 234
Criots-Bâtard-Montrachet 34, 146, 234, 246
La Croix Blanche 137
Les Croix Moines 255
'cru' definition 299
cuisine and wine 272–3
Curley, Róisín 232
Cuvérie Corton Grancy 175

Les Damodes 138
Les Demoiselles 229
Devauges, Jacques 97
Dijon 10–11, 13, 20
Domaine & Maison Marchand-Tawse 152–4
Domaine Alain Gras 192–3
Domaine Alain Hudelot-Noëllat 118–19
Domaine Alex Moreau 236–7
Domaine Anne Gros 123–4, 284
Domaine Arlaud 91–3

Domaine Armand Heitz 235–6
Domaine Armelle et Bernard Rion 137–8
Domaine Bachey-Legros 248–9
Domaine Bertagna 115–16
Domaine Bitouzet-Prieur 199–201
Domaine Bonneau du Martray 160–62
Domaine Bouchard Père & Fils 168–70
Domaine Bruno Clair 68–69
Domaine Bruno Colin 230–31
Domaine Cécile Tremblay 103–4
Domaine Chandon de Briailles 164–5
Domaine Chanson Père & Fils 170–72
Domaine Charles Audoin 67–8
Domaine Chevrot et Fils 254–5
Domaine Claude Dugat 79–80
Domaine Comte Georges de Vogüé 110–12
Domaine de la Pousse d'Or 188–9
Domaine de la Romanée-Conti 27, 138–142,
  160
Domaine de la Vougeraie 153, 154–6
Domaine de Montille 213–14
Domaine des Comte Lafon 205–6
Domaine des Lambrays 97–8
Domaine des Terres de Velle 195
Domaine du Château Philippe le Hardi 251–2
Domaine du Clos Tart 102–3
Domaine du Comte Armand – Clos des
  Epeneaux 180–82
Domaine du Comte Liger-Belair 128–30
Domaine Duc de Magenta 154
Domaine Dujac 19, 93–5, 284
Domaine Duroché 180–81
Domaine Etienne Sauzet 224–5
Domaine Evenstad 249–50
Domaine Fabien Coche 201–2
Domaine Faiveley 146–8, 284
Domaine Fontaine-Gagnard 233–4
Domaine Fourrier 83–4
Domaine François Carillon 218–19
Domaine Georges Lignier et Fils 99–100
Domaine Georges Roumier 109–10
Domaine Gros Frère et Soeur 125–6
Domaine Guy Amiot et Fils 228–9
Domaine Henri & Gilles Buisson 191–2
Domaine Henri Gouges 148–50
Domaine Henri Rebourseau 84–5
Domaine Hubert Lamy 246–7
Domaine Hubert Lignier 100–101
Domaine Jacques Carillon 219–20
Domaine Jacques-Frédéric Mugnier 107–8
Domaine Jean Chartron 221–2
Domaine Jean-Claude Bachelet et Fils 243–4
Domaine Jean-Claude Ramonet 239–42
Domaine Jean-Marc Vincent 252–3
Domaine Latour-Giraud 208–9

INDEX 309

Domaine Leflaive 222–4, 284
Domaine Les Astrelles 76–7
Domaine Marc Colin 244–5
Domaine Marc Morey 238–9
Domaine Marquis d'Angerville 185–7
Domaine Matrot 210–11
Domaine Méo-Camuzet 131–2, 284
Domaine Michel Lafarge 187–8
Domaine Michel Noëllat 135–6
Domaine Nicole Lamarche 127–8
Domaine Patrick Javillier 204–5
Domaine Philippe Charlopin 77–9
Domaine Pierre Gelin 72–3
Domaine Pierre Labet 120
Domaine Pierre Vincent 195–6
Domaine Ponsot 23, 53
Domaine Rapet Père et Fils 162–3
Domaine Robert Ampeau et Fils 198–9
Domaine Robert Groffier Père et Fils 95–6
Domaine Rossignol-Trapet 86–7
Domaine Roulot 215–16
Domaine Seguin-Manuel 178–9
Domaine Serafin Père et Fils 87–9
Domaine Sylvain Pataille 70–71
Domaine Sylvie Esmonin 82–3
Domaine Thibault Liger-Belair 150–52
Domaine Tollot-Beaut 165–7
Domaine Trapet Père et Fils 89–90
Domaine Vincent Latour 207–8
Domaine Vincent & Sophie Morey 239–40
DRC 27, 138–142, 160
Drouhin family 172
Drouhin, Joseph 5, 284
Drouhin, Véronique 172–3
du Dessus 136
Dugat family 79
Dupard, Jean-Edouard 221
Duroché, Pierre and Marianne 80–81

Les Echéseaux 137
Echézeaux **19**, 26, 79, 104, 127–8, 130, 136, 295
  lieux-dits 28, 126
  producers of 133, 134
  vineyard 123
  wines 134, 140, 141
    under production 94, 133
  see also Flagey-Echézeaux; Grands Echézeaux
élevage 62
Les Embrazées 240
En la Rue de Vergey 23
En Montceau 245–6
En Remilly 36, *227*, 245, 246
Es Chézot 20
Evenstad, Ken and Grace 249–50

Les Evocelles 21

Faiveley family 146–8, 284
Fenal, Perrine 139
fermentation
  and *élevage* 62
  of red wine 60–61
  of white wine 62
  by domaine *see* individual names of domaines
fertilizers
  artificial 16–17, 41
  natural 182
Les Feusselottes 250–51
fining and filtering 61
Fixey 71–2
Fixin 20–21, 29, *65*, 71–3
Flagey-Echézeaux 121, *122*, 123
  see also Echézeaux; Les Grands Echézeaux
Follin-Arbelet, Stéphane 212
Fontaine, Céline 234
food and wine 272–3
Fourrier, Jean-Marie 83–4, 284
fraud 14, 282
French Revolution 11–12, 285
Les Fuées 112

Gagey family 174
Gamay grape (Bourgogne Passetoutgrains) 53, 72, 90–91, 97, 124–5, 188
La Garenne 224
Gelin, Pierre-Emmanuel 72–3
Les Genevrières 33, 209
Germain, Eric 203
Gevrey 72, 73
Gevrey-Chambertin 4
  history 5
  producers of 76–91
  topology 37, *65*, 296
  vineyards and holdings 21, 68–9, 72, 251
    maps of *74–5*
  wines 73, 76, 78, 99, 101, 120, 127
    grands crus 85, 87
    premiers crus **19**, 80, 82, 83, 84, 89, 252
    village 82
Girardin, Vincent 202–3
Giraud, Bastien 85
glassware 272
Gouges, Grégory and Antoine 148–9
Gouges, Henri 15
Goulley, Frédérique 87–8
Les Grand Jours de Bourgogne 230, 290–91
La Grande Rue 15, 27, *122*, 128, 129
Les Grands Echézeaux 26–7, 28, 123, 137, 140
  see also Flagey- Echézeaux; Grands Echézeaux
Grands Epenots **33**, *181*, 182

310 CÔTE D'OR

Gras, Alain and Arthur 191–2
Les Gravières 36, 190, 240
   premiers crus 231–2, 258
   vineyard *36*
Les Grèves 30, 167, 212
Groffier, Nicolas 95–6
Gros family 123, 284

Haisma, Mark 117–18
harvesting 51–2, 222
   regional 53–6
   by year *see* vintage reputation
Hautes-Côtes de Nuits *65,* 136, 151
Healy, Maurice 90
Heitz, Armand 235–6
Henriot, Joseph 169
herbicides 16–17, 276
Les Héritiers de Louis Jadot 174
Les Hervelets 20–21
Hill of Corton
   history 7
   producers 161–5, 191
   topography 38, *158,* 296
   wines and planting 29–30, 157–8
   see also Aloxe-Corton; Corton; Corton-
      Charlemagne; Charlemagne
Hospices de Beaune wine auction 293
Hôtel-Dieu, Beaune 292
Hyde family 284

inheritance tax 281

Jacquet, Thibault 160
Jadot family 174, 284
Les Jarolières 31, *181*
Javillier, Marion 204–5
Jayer, Henri 59

labelling 4, 30
Labet, François and Edouard 120
Ladoix-Serrigny 29, 37, 148, 157, 159–60
Lafarge, Frédéric and family 187–8, 284
Lafon, Dominique 205
Lafon, Pierre and Léa 205–6
Lamarche, Nicole and Nathalie 127–8, 129
Lamy, Pierre-Emmanuel 204
Lamy, Olivier 246–7
land ownership
   headline sales 283
   legislative hamstringing 287
   and migration 283–4
   purchase prices 281–2
Landanger family 188
Latour, Eléonore 176

Latour, Jean-Pierre and Josephine 208–9
Latour, Louis-Fabrice 175–6
Latour, Vincent and Cécile 207–8
Latricières-Chambertin 87
Lavalle, Dr Jules 12–13
Lavaux Saint-Jacques 80, 82
Les Lavières 166
le Bon, Phillippe 10
le Hardi, Philippe 10–11
Le Tesson 33, *197*
Lecoeur, Arnaud 115
Legros, Lenaïc and Samuel 248–9
Leprince, Frédéric 193–4
Leroux, Benjamin 177–8, 194
lieu-dit labelling 30
lieu-dit/climat 24
Liger-Belair, Thibault 150–52, *152,* 284
Liger-Belair, Vicomte Louis-Michel 128–30
Lignier, Laurent and Sébastien 100
Les Loächausses 126
Lupatelli, Jean 110–12

Mâconnais 284
Magnat, Marc 250
Maison Benjamin Leroux 177–8
Maison de Montille 213
Maison François Millet & Fils 105–7
Maison Frédéric Leprince 193–4
Maison Jean-Claude Boisset 144–6
Maison Joseph Drouhin 5, 172–3, 284, 294
Maison Louis Jadot 173–5
Maison Louis Latour 175–7, 284
Maison Mark Haisma 117–18
Maison Róisín Curley 232–3
La Maison Vougeot 295
Maison Vincent Girardin 202–3
Malconsorts 28, 119–20, *122,* 128, 130–31
La Maltroie 231
Maranges
   appellation 254
   premier cru 255
   producers 254–5
   topography 39, *158*
   vineyards and holdings 36, 239, 248
   vintage reputation 260
Marchand-Tawse 25, 152–3
Marchand, Pascal 152–4
Marsanny-la-Côte 20, 64, *65*
   producers 66–71, 78
Mathiaud, Cécile *ix*
Matrot, Elsa and Adele 210
Mazis-Chambertin 22, *74,* 77, 78, 85–6
Meo, Jean-Nicolas 131, 284
merchants *see* négociants

INDEX 311

Meursault 159
  renown 196–8
  topography 38–9, *158,* 190, 196
  vineyards 32–3, 182
    map of *197*
  vintage reputation 261–2
  wines 120, 196
    premier cru 177–8, 203, 208, 209, 213,
      236
    under production 172, 192–3
    villages 120, 202, 205, 236
Millet, François 105–7
Millet, Julien 76
'minerality' **52**
Mollard, Sabine 238
monasteries 6–9, 11
Monthelie
  producers 100
  topography 157, *158*
  wines 190, 195, 204
Montille, Alix de 213
Montille, Etienne de 194, 213, 223, 284
Montille, Hubert de 213
Montrachet
  satellites *see* Bâtard-Montrachet; Bienvenues-
    Bâtard-Montrachet; Chevalier-
    Montrachet; Criots-Bâtard-Montrachet
  topography 33–4, 39
  wine 234
Monts Luisants 23, 24, 53, *92,* 101
Morandière, Brice de la 222–3
Moreau, Alex 236–7
Morelot, Denis-Blaise 12
Morey-Saint-Denis 8
  premiers crus 109, 118
  topography 23, *65,* 91, *113*
  vineyards and holdings 23, 53, 69, 88
    maps of *92*
  visiting 291
Morey, Sylvian 284
Morgeot 35, 235–6, 240, 242, 249
Mugnier, Frédéric 107–8
Les Murgers des Dents de Chien 36
Murat, Emilie *ix*
Musigny 24–6, *34, 64*
  lieux-dits and plantings 25
  renown 25–6
  vineyards and holdings 24–6, *43,* 108
  wines 110, 112
    grand cru 148, 154
    under production 111–12
Les Musigny 25

Nantoux 193–4

Les Narvaux 33, *197*
négociants 31, 102, 152, 168, 177
  business model changes 275–6
  fight over origin 15–16
  micro-négocients **285**
Negrerie, Paul 183
Nicolay-Jousset, Claude de 164
Nicolay, François de 164
Noëllat, Sophie and Sébastien 135–6
Noli, Alessandro 102–3
Nuits 142
Nuits-Saint-Georges
  history 15
  producers 144–56
  topography 142, *143,* 162
  wines 127, 144, 154, 179, 180
    premier cru 98–9, 108–9, 136, 138, 149,
      150, 151
Nuits-Saints-George Les Longcourts 250

oak 47, 51, **61,** 114, 204, 273
organic vineyards 42–4, 69, 71–2, 90, 97, 109,
  115, 131, 147, 154, 178, 251–2, 255

Pataille, Sylvain 20, 70–71
Patriat, Grégory 144–6
Paulée de Meursault 212
Les Perrières, Beaune *premier cru* 176–7
Les Perrières, Meursault *premier cru* 33, 198–9,
  203–4, 207, 236
Les Perrières, Nuit-Saints-George *premier cru*
  143, 153
Les Perrières, Pommard vineyard *181*
Les Perrières, Puligny-Montrachet *premier cru*
  217, 220
Pernand-Vergelesses
  producers 160–63
  topology 29, 157–8, *158,* 160, 296
  wines 241
    premiers crus 117, 176
    villages 117
pesticides 276
Les Petits Clos Vieilles Vignes 249
Petits Epenots **33,** *181,* 182
Les Petits Musigny 25
phylloxera 13–14, 17, 41
pickers 55, *166*
pigeage and *remontage* 60
Pinot Gouges grape 149
Pinot Noir grape 48–50
  Passetoutgrains blend 53, 72, 90–91, 97,
    124–5, 188
  under production 66, 90–91, 105, 127, 186
Les Pins 117

312 CÔTE D'OR

Pitiot, Sylvain 297
Les Plantes Momières 248
Pommard 4, **33**
  producers 100, 180–84
  topography 31, 180, *181*
  wines 159
    premiers crus 77, 182, 214, 236
    villages 179, 183–4, 189, 206
Les Porlottes 112, *113*
Les Porroux 101
Potel, Gérard 188
Les Pougets 162
Poupon, Pierre 297
Le Pré de Manche 201
premox (premature oxidation) 62–3, 222,
  276–8
prices of wine at auction 281–2
producers *see* individual names of Domaine
Prohibition **286**
Provence, Joël *x*
Les Pucelles 35, *217,* 223–4
Puligny-Montrachet
  producers 77, 218–25, 248
  renown 216–17
  Saint-Vincent Tournante *290*
  topography 39
  vineyards and holdings 195, 200
    map of *217*
  wines 33, 35, 159, 216–17, 222, 248
    premiers crus 189, 195, 199, 206, 210,
    220, 223–4, 229, 245
    villages 196, 219
purchase price of land 281

racking 61
Ramonet family 240–42
Rapet, Vincent 162
red wine
  of Auxey-Duresses 199
  Bourgogne Rouge 119, 131, 151, 202
  of Chassagne-Montrachet **50,** 240, 258
  Côte de Beaune 206, 214
  of Marsannay-la-Côte 70
  of Puligny 35
  vinification 58–61
  vintage reputation (2024–1985) 258–70
    older vintages of note 269–70
  Volnay 189
  see also Gamay grape; Pinot Noir grape;
    winemaking, red wine
regulation 6, 12, 14–16
Reh, Eva 115
Les Renardes 30
Richebourg 131–2
  topography *122*

vineyard holdings 140–41
wine 27–8, 119, 123–4, 126, 131–2, 141,
  151
Riffault, Benoît 224
Rion family 137
ripeness 54
La Romanée
  history 11
  planting 27
  topography *122,* 130, *226*
  vineyard ownership 129–30
  wine 130
    premier cru 203, 235
La Romanée-Conti
  history 11
  planting 130
  vineyard 16, 27, *38*
  wine 59, 123, 141
Romanée-Saint-Vivant 27–8, 125, 140–41
remontage 60
rosé 137
Rossignol, Nicolas and David 86
Roulot, Jean-Marc and Félicien 215
Roumier, Christophe 109–10
Les Ruchots 92
Les Rugiens 31, *181*

Saint-Aubin 192
  producers 44, 242–7
  renown 242
  topography 157, *158, 227,* 242
  wine 207, 242
    premiers crus 244, 245, 247
Saint-Romain
  producers 191–3
  topography 32, 157, *158,* 190
  vineyard holdings 173
  wine 192, 193, 232
Saint-Vincent Tournante 289–90
Les Saints-Georges 28, 142, 149, 151
La Salle de la Paulée 212
Santenay
  premier cru 190, 231
  producers 240, 248–54
  renown 247–8
  topography 39, *158,* 247
  vineyards *36*
Les Santenots du Milieu 32, *184, 197*
Santenots (Meursault) 32, *197*
  premier 200
Santenots (Volnay) 32,
  wine 199
    premiers crus 203, 206
Savigny-lès-Beaune
  premier cru 165

INDEX   313

producers 164–5, 169
topography 30, 157, *158*
vineyard holdings 68, 195
Seiter, Thomas 174
Les Serpens 207–8
serving wine 271–3
Seysses, Jeremy and Diana 94, *95*
Sieve, Brian 214
Sirugue, Arnaud 136
Les Sizies 214
smallholders, emergence of 10–11
Soler'Al 253–4
sorting 58, 224, 287
Sous le Château 92
Stehly, Benoît 99–100
stems 59, 103
    practices in Beaune 171, 179
    practices in Chambolle-Musigny 107, 109,
        119,
    practices in Gevrey-Chambertin 86, 88
    practices in Meursault 200, 203, 214
    practices in Morey-Saint-Denis 100, 101,
        103, 104
    practices in Nuits-Saint Georges 153
    practices in Puligny-Montrachet 224
    practices in Saint Romain 192
    practices in Savigny-lès-Beaune 171, 179
    practices in Volnay 187
    practices in Vosne-Romanée 131, 133, 134
Les Suchots 28, *122,* 127, 128, 130, 136
sulphur levels 278

La Tâche 27, *122,* 123, 129, 141
Taille Pieds (Taillepieds) 32, *184,* 201
Tawse, Morey 153
temperance movement **286**
terroir messages 40–41, 54
terroir, definitions 18–20, 24
Tête de Murger 204
Thibaut, Marion 178–9
Les Tillets 33, 120, *197,* 205
Tollot-Gros, Paul 123–4
Tollot, Jean-Paul 284
Tollot, Nathalie 165–6
Trapet, Jean-Louis 89–90
Tremblay, Cécile 103–4

UNESCO 294–5
Les Ursulines 144–6

Valois dukes 9–10
Les Valozières 56
van Canneyt, Charles 118–19, 134
Les Vaucrains 150
Les Vaumuriens 206

Les Vergers 234
Les Vérroilles 126
vignerons *see* viticulture *and* individual names
    of domaines
'village'/*village* 3–4
Villaine, Aubert de *ix–x,* 139
Villaine, Bertrand de 139, 284
Vincent, Jean-Marc *ix,* 252–3
Vincent, Pierre 195–6, 223
vineyard to winery 47–8
vineyards
    biodynamic 42–4, 90, 164, 178, 224, 243
    bougies (smudge pots) 279, *280*
    clos designated *see* individual names of Clos
    fragmentation 283
    geologically most favoured 40
    mobile brazier *43*
    names **33**
    organic *see* organic vineyards
    planting pre- and post-phylloxera 41
    purchase prices 281
    ranking 44
    tow and blow trailer *257*
    visiting 296–7
    see also viticulture
vinification *see* winemaking
vintages
    assessment changes and pigeonholing
        257–8
    spring frost and summer hail 256–7
    (2024–1985) 259–68, 276
        older vintages of note 269–70
Les Vireuils 33, *197*
viticulture 41–4
    future 286–8
    post-1985 275–6
    terroir 18–20, 24, 40–41, 54
    see also vineyards
Volnay
    history 8, 14–15, 168
    producers 185–90, 213
    topography 38, 184–5
    vineyards and holdings 31–2, 100, 173, 200
        map of *184*
    wines 28, 185, 188
        premiers crus 32, 107, 176, 178, 186–7,
            201, 206–7, 209, 214
        Santenots 199, 203
        villages 194–5, 199
Vosne-Romanée
    producers 123–42
    topography 38, *38, 65*
        map of vineyards *122*
    vineyards and holdings 4, 27–8, *38,* 69, 73,
        115, 122–3

314 CÔTE D'OR

visiting 295–6
wines 151
  grands crus 28
  premiers crus 79, 98, 119–20, 130, 136, 146, 178
  village 104, 124–5, 179
Vougeot 8, *113*, 114, 115–21, 155, 305

weather *see* climate and weather
white wine
  Bourgogne Blanc 149, 214, 216, 223, 236, 237
  foremost *see* Chassagne-Montrachet; Meursault; Puligny-Montrachet
  grapes *see* Aligote grape; Chardonnay
  vintage reputation (2024–1985) 258–70
    older vintages of note 269–70
  winemaking *see* winemaking, white wine
winemaking
  premox 62–3, 76–9, 222
  producers' methods *see* individual names of Domaine
  sulphur levels 278

see also winemaking, red wine; winemaking, white wine
winemaking, red wine
  chaptalization/adding sugar 60–61
  cool soak 60
  fermentation 60–61
  fining and filtering 61
  marc 61
  pigeage and *remontage* 60
  racking 61
  sorting 58
  stems and 59
  tartaric acid 61
  vessels 61
  see also red wine; winemaking
winemaking, white wine 61–3
  acidity 62
  bâtonnage 62
  oxidation and bottling closures 62–3
  from sorting to press 62
  see also white wine; winemaking

Zinetti, Paul 181–2

More from The Classic Wine Library

The Classic Wine Library is a premium source of information for students of wine, sommeliers and others who work in the wine industry, but can easily be enjoyed by anybody with an enthusiasm for wine. All authors are expert in their subject, with years of experience in the wine industry, and many are Masters of Wine. The series is curated by an editorial board made up of Sarah Jane Evans MW, Richard Mayson and James Tidwell MS. Continue expanding your knowledge with the titles below.

The wines of Chablis and the Grand Auxerrois

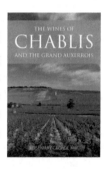

Features the history, vineyards, *crus* and viticultural methods of Burgundy's most northerly wine-growing region, profiling the producers who make Chablis in the twenty-first century and assessing what the future holds for these historic wines. The book also explores the wines and producers of the neighbouring wine region, the Grand Auxerrois.

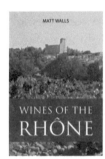

Wines of the Rhône

This guide to one of the great French wine regions covers all the appellations of the Rhône, featuring interviews with some of the most respected winemakers of the region, and tackles the issues facing the Rhône's wines with clarity and authority. An ideal introduction for those new to the Rhône, while providing fresh insights for long-time admirers of the wines.

Wines of the Loire Valley

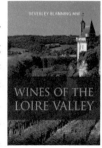

The wines of this historic region are renowned for their balance between succulent fruit and refreshing acidity. The book begins with a history of wine in the Loire valley and provides an overview of the region's terroir, grape varieties and wine styles. The regional section of the book covers the famous appellations and producers, as well as many that are overlooked and undervalued, and highlights some hidden gems.

Browse the full list and buy online at
academieduvinlibrary.com

Also published by Académie du Vin Library

We choose our books with care – above all for their readability, but also because we genuinely believe they have something important to say about the world of fine wine that will enhance your drinking pleasure. We are proud to publish many of the world's leading authors and wine experts.

On Burgundy: From maddening to marvellous in 59 wine tales

Burgundy can baffle with its unwillingness to charm in the glass, then dazzle and astound us when we catch it at its best. Those who know burgundy frequently never love another wine. The authors and experts in this book answer the questions everyone asks: Why do these wines taste as good as they do? Who makes them this way?

Taste the Limestone, Smell the Slate
Alex Maltman

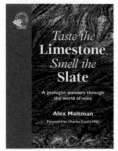

Alex Maltman analyses the link between vineyard geology and wine flavour. Highlighting connections that many of us are simply oblivious to, he outlines the science, unravels the metaphors and challenges us to question the accepted truths we read on wine labels and in the press.

Behind the Glass: The chemical and sensorial terroir of wine tasting
Gus Zhu MW

What makes red wine red? Do genetic differences, culture and life experience change our perception of wine? And is there science behind the obscure language of tasting notes? The answers to all of these questions and more are explored in *Behind the Glass*, a smart, accessible investigation into the science behind a glass of wine – and our appreciation of it.

Browse the full list and buy online at
academieduvinlibrary.com